# ANATOMY & PHYSIOLOGY II

## BIO 169-900R

*Forsyth Tech
Community College*

*Human Biology
Department*

# *Anatomy and Physiology II*

BIO 169-900R
Forsyth Tech Community College
Human Biology Department



Printed in the United States of America
10 9 8 7 6 5 4 3
ISBN: 978-1-61740-805-2

Van-Griner Publishing
Cincinnati, Ohio
www.van-griner.com

President: Dreis Van Landuyt
Project Manager: Maria Walterbusch
Customer Care Lead: Lauren Houseworth

Rudolph 805-2 Su19
314192-319326

# TABLE OF CONTENTS

## LABORATORIES

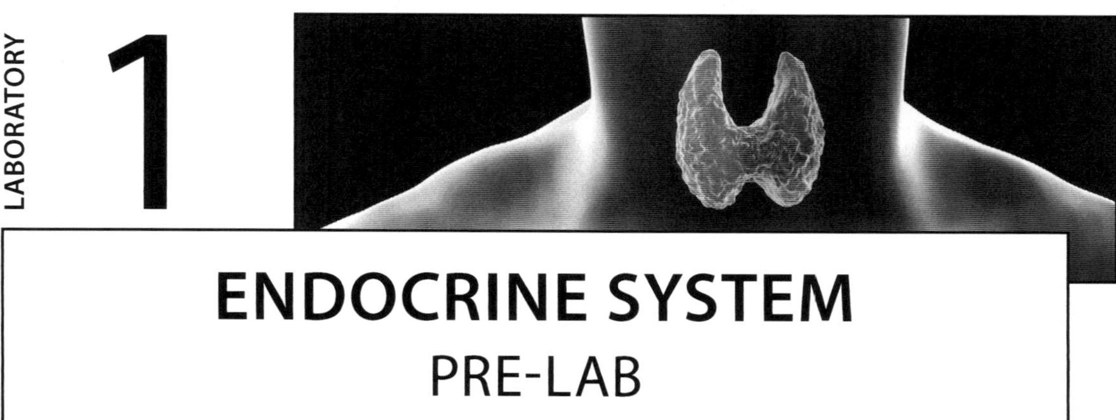

# ENDOCRINE SYSTEM
## PRE-LAB

Name: _________________________  Section: __________  Date: __________

## LEARNING OBJECTIVES

1. Identify and characterize the two most important systems for maintaining homeostasis.

2. Identify and locate the glands that make up the endocrine system.

3. Detail the general functions of the hormones secreted by each endocrine gland.

4. Discuss the relationship between the hypothalamus and pituitary gland.

*Checklist to complete* **before entering** *the science skills lab (SSL):*

☐ Actively read this packet of information.

☐ Complete the charts, tables or labeling and answer questions using your own words.

☐ Complete the electronic digital pre-lab quiz on Blackboard (Bb) by Sunday.

☐ Complete the Blood Glucose Regulation Wiley PowerPhys #5 via Bb.

☐ Review the attached anatomy list and take it to lab with you. Jot down key descriptive identifying words that aid in your lab test preparations.

You will spend **2 hours or so** in lab at Forsyth Tech to complete the following activities. This amount of time allows you to complete the activities by using the torso model, using other models. and working with a lab partner.

# ACTIVITY 1: COMPARE AND CONTRAST THE NERVOUS AND ENDOCRINE SYSTEMS

*Read the content in your text and the introductory paragraphs below. Then complete Activity 1 in lab.*

Communication is a process in which a sender transmits signals to one or more receivers to control and coordinate actions. In the human body, two major organ systems participate in relatively "long distance" communication: the nervous system and the endocrine system. Together, these two systems are primarily responsible for maintaining homeostasis in the body.

So how are these two systems similar? How do they differ? Both systems utilize chemical messengers but differ in the way that the message is delivered. The nervous system communicates directly with its target via short distance chemical messengers called neurotransmitters. Due to the direct nature of communication, neurotransmitters are released very near their targets after an action potential arrives at the synaptic end bulb. As a result, neurotransmitters act very rapidly on local targets (receptors). The effect of a neurotransmitter is short lived because once the neurotransmitter has been released, it can be destroyed by enzymes, diffuse away, or even be taken back up by the secreting cell. In this way, neural communication enables body functions that involve quick, brief actions, such as movement, sensation, and cognition.

In contrast, the *endocrine* system communicates indirectly with its targets via long distance chemical signals. The signal communication is known as indirect because the signal travels throughout the bloodstream and eventually ends up at the target. These chemical signals, known as hormones, are created by specialized cells that comprise endocrine glands. Once the stimulus for hormone release is received, the hormone is sent into the circulatory system and travels throughout the body before reaching its target and inducing a response. As a result, endocrine signaling requires more time than neural signaling to prompt a response in target cells, though the precise amount of time varies with different hormones.

In addition, endocrine signaling is typically less specific than neural signaling. The same hormone may play a role in a variety of different physiological processes depending on the target cells involved. For example, the hormone oxytocin promotes uterine contractions in women in labor. It is also important in breastfeeding and may be involved in the sexual response and in feelings of emotional attachment in both males and females.

## ACTIVITY 2: IDENTIFY AND LOCATE ALL ENDOCRINE GLANDS

*Read the content in your text and the introductory paragraphs below. Then complete Activity 2 in lab.*

The **endocrine system** consists of cells, tissues, and organs that secrete hormones as a primary or secondary function. The endocrine glands are the major players in this system. The primary function of these ductless glands is to secrete their hormones directly into the surrounding interstitial fluid. The interstitial fluid and the blood vessels then transport the hormones throughout the body. Notice that this is a two-step process. The endocrine glandular cells secrete the hormone into the surrounding interstitial fluid. The hormone must then travel from the interstitial fluid into the bloodstream before it can be carried to its target. The endocrine system includes the **ovaries, testes, thymus, hypothalamus, pancreas, pituitary, thyroid, parathyroid, adrenal, and pineal glands.**

## ACTIVITY 3: EXAMINE THE RELATIONSHIP BETWEEN THE HYPOTHALAMUS AND PITUITARY GLAND

*Read the content in your text and the introductory paragraphs below. Complete the table below, and then complete Activity 3 in lab.*

The nervous and endocrine systems work together to maintain homeostasis. The hypothalamus serves as the link between the two systems and allows stimuli received by the nervous system to stimulate the endocrine system and bring about long-term homeostasis.

The **hypothalamus** is a structure of the diencephalon of the brain, located anterior and inferior to the thalamus. It has both neural and endocrine functions, producing and secreting many hormones. In addition, the hypothalamus is anatomically and functionally related to the **pituitary gland** (or hypophysis), a bean-sized organ suspended from it by a stem called the **infundibulum** (or pituitary stalk). The pituitary gland is cradled within the sella turcica of the sphenoid bone of the skull. It consists of two lobes that arise from distinct parts of embryonic tissue; the posterior pituitary (neurohypophysis) is neural tissue, whereas the anterior pituitary (also known as the adenohypophysis) is glandular tissue that develops from the primitive digestive tract.

The anterior pituitary known as the **adenohypophysis** (adeno- = gland) is made out of glandular cells that manufacture and secrete hormones. The secretion of hormones from the anterior pituitary is regulated by two classes of hormones. These hormones—secreted by the hypothalamus—are the releasing hormones that stimulate the secretion of hormones from the anterior pituitary and the inhibiting hormones that inhibit secretion. Hypothalamic hormones enter the anterior pituitary through blood vessels.

Within the infundibulum is a bridge of capillaries that connects the hypothalamus to the anterior pituitary. This network, called the **hypothalamic—hypophyseal portal system,** allows hypothalamic hormones to be transported to the anterior pituitary without first entering the systemic circulation. Hormones from the adenohypophysis include **human growth hormone (hGH), adrenocorticotropic hormone (ACTH), thyroid stimulating hormone (TSH), follicle stimulating hormone (FSH), luteinizing hormone (LH), prolactin (PRL),** and **melanocyte stimulating hormone (MSH).**

The posterior pituitary known as the **neurohypophysis** is made out of neural tissue (neuro- = neural). The dendrites and cell bodies of the neurons rest in the hypothalamus, but their axons descend as the hypothalamic–hypophyseal tract within the infundibulum and end in axon terminals that comprise the posterior pituitary. This means that the posterior pituitary gland does not produce hormones but rather, stores and secretes hormones produced by the hypothalamus. These hormones travel along the axons into storage sites in the axon terminals of the posterior pituitary. In response to signals from the same hypothalamic neurons, the hormones are released from the axon terminals into the bloodstream. Hormones from the neurohypophysis include **antidiuretic hormone (ADH)** and **oxytocin (OT).**

| ENDOCRINE GLAND | HORMONE RELEASED (SPELL OUT FULL NAME) |
| --- | --- |
| Anterior Pituitary (Adenohypophysis) | |
| | |
| | |
| | |
| | |
| | |
| | |
| Posterior Pituitary (Neurohypophysis) | |
| | |

## ACTIVITY 4: DETAIL THE GENERAL FUNCTIONS OF THE HORMONES SECRETED BY EACH OF THE ENDOCRINE GLANDS

*Read the content in your text and the introductory paragraphs below. Complete the table below. Then, complete Activity 4 in lab.*

### THYROID GLAND

A butterfly-shaped organ, the thyroid gland is located in the neck, anterior to the trachea, just inferior to the larynx. The medial region, called the **isthmus,** is flanked by wing-shaped left and right **lobes.** Each of the thyroid lobes are embedded on their posterior surfaces with smaller, pea-sized, parathyroid glands. The glandular tissue of the thyroid gland is composed mostly of thyroid follicles. The follicles are made up of a central cavity filled with a sticky fluid called **colloid.** Surrounded by a wall of epithelial follicle cells, the colloid is the center of thyroid hormone production, and that production is dependent on the supply of the hormones' essential and unique component, iodine. When iodine is present, the **follicular cells** manufacture **T3 and T4.** Surrounding the follicles are larger **parafollicular cells** also known as C cells. These cells do not participate in the formation of T3 and T4 but rather make their own hormone, **calcitonin.**

### PARATHYROID GLANDS

The parathyroid glands are tiny, round structures usually found embedded in the posterior surface of the thyroid gland. A thick connective tissue capsule separates the glands from the thyroid tissue. Most people have four parathyroid glands, but occasionally, there are more in tissues of the neck or chest. The primary functional cells of the parathyroid glands are the **principle cells.** These epithelial cells produce and secrete the **parathyroid hormone (PTH),** the major hormone involved in the regulation of blood calcium levels. PTH is secreted in response to low blood calcium levels and primarily targets the bones and kidneys.

### ADRENAL GLANDS

The adrenal glands are wedges of glandular and neuroendocrine tissue adhering to the top of the kidneys by a fibrous capsule. The adrenal gland consists of an outer cortex of glandular tissue and an inner medulla of nervous tissue. The cortex itself is divided into three zones: the **zona glomerulosa,** the **zona fasciculata,** and the **zona reticularis.** Each region secretes its own set of hormones. The **adrenal cortex** secretes steroid hormones which are important for the regulation of the long-term stress response, blood pressure and blood volume, nutrient uptake and storage, fluid and electrolyte balance, and inflammation.

The **adrenal medulla** is part of the fight or flight response by being an extension of the sympathetic autonomic nervous system. The sympathomedullary (SAM) pathway involves the stimulation of the medulla by impulses from the hypothalamus via neurons from the thoracic spinal cord. The medulla is stimulated to secrete the catecholamine hormones **epinephrine and norepinephrine.**

## PINEAL GLAND

Recall that the hypothalamus, part of the diencephalon of the brain, sits inferior and somewhat anterior to the thalamus. Posterior to the thalamus is the **pineal gland,** a tiny endocrine gland whose functions are not entirely clear. The pinealocyte cells that make up the pineal gland are known to produce and secrete the amine hormone **melatonin,** which is derived from serotonin. The secretion of melatonin varies according to the level of light received from the environment. When light stimulates the retinas of the eyes, the production of melatonin is inhibited. As a result, blood levels of melatonin fall, promoting wakefulness. In contrast, as light levels decline—such as during the evening—melatonin production increases, boosting blood levels and causing drowsiness.

## PANCREAS

The pancreas is a long, slender organ, most of which is located posterior to the bottom half of the stomach. Although it is primarily an exocrine gland, secreting a variety of digestive enzymes, the pancreas has an endocrine function. Its **pancreatic islets**—clusters of cells formerly known as the islets of Langerhans—secrete the hormones **glucagon, insulin,** somatostatin, and pancreatic polypeptide (PP). The hormones glucagon and insulin target muscle cells and liver cells to regulate the amount of glucose in the blood.

## TESTES AND OVARIES

The primary hormone produced by the male testes is **testosterone,** a steroid hormone important in the development of the male reproductive system, the maturation of sperm cells, and the development of male secondary sex characteristics such as a deepened voice, body hair, and increased muscle mass. In addition, the testes produce the peptide hormone **inhibin,** which inhibits the secretion of FSH from the anterior pituitary gland. FSH stimulates spermatogenesis.

The primary hormones produced by the ovaries are **estrogens,** which include estradiol, estriol, and estrone. Estrogens play an important role in a larger number of physiological processes, including the development of the female reproductive system, regulation of the menstrual cycle, the development of female secondary sex characteristics such as increased adipose tissue and the development of breast tissue, and the maintenance of pregnancy. Another significant ovarian hormone is **progesterone,** which contributes to regulation of the menstrual cycle and is important in preparing the body for pregnancy as well as maintaining pregnancy. In addition, the granulosa cells of the ovarian follicles produce inhibin, which—as in males—inhibits the secretion of FSH.

| ENDOCRINE GLAND | HORMONE RELEASED (SPELL OUT FULL NAME) |
| --- | --- |
| Thyroid | |
| | |
| Parathyroid | |
| Adrenal | |
| | |
| | |
| Pineal | |
| Pancreas | |
| | |
| Testes | |
| Ovaries | |

*Checklist to complete* **before entering** *the science skills lab (SSL):*

☐ Actively read this packet of information.

☐ Complete the charts, tables or labeling and answer questions using your own words.

☐ Complete the electronic digital pre-lab quiz on Bb by Sunday.

☐ Complete the Blood Glucose Regulation Wiley PowerPhys #5 via Bb.

☐ Review the attached anatomy list and take it to lab with you. Jot down key descriptive identifying words that aid in your lab test preparations.

## ANATOMY LIST

1. Identify the following structures associated with the endocrine system:
   a. sella turcica of the sphenoid bone
   b. hypothalamus
   c. infundibulum
   d. pituitary gland (hypophysis)
   e. pineal gland (pineal body)
2. Identify the following microscopic structures on a slide of the pituitary gland:
   a. anterior lobe of the pituitary gland (adenohypophysis) (pars distalis)
   b. posterior lobe of the pituitary gland (neurohypophysis) (pars nervosa)
3. Identify the following structures associated with the endocrine system:
   a. thyroid gland
      i. right lobe
      ii. left lobe
      iii. isthmus
   b. parathyroid glands
4. Identify the following microscopic structures of the thyroid and parathyroid glands:
   a. thyroid follicles
      i. follicular cells
      ii. follicular cavity with thryroglobulin (for physiology)
   b. extrafollicular cells (parafollicular cells) ( C cells)
   c. parathyroid gland
5. Identify the following structures associated with the endocrine system models/ diagrams:
   a. adrenal (suprarenal) glands
6. Identify the following microscopic structures on a slide of the adrenal gland:
   a. adrenal capsule
   b. adrenal cortex
      i. zona glomerulosa
      ii. zona fasciculata
      iii. zona reticularis
   c. adrenal medulla

7. Identify the following structures associated with the endocrine system models/diagrams:

    a. pancreas

        i. head

        ii. body

        iii. tail

        iv. primary (main) pancreatic duct (duct of Wirsung)

    b. common bile duct

8. Identify the following microscopic structures on a slide of the pancreas:

    a. pancreatic acini ( contain pancreatic acinar cells)—exocrine portion

    b. pancreatic islets (Islets of Langerhans)

9. Identify the following endocrine structures on the appropriate models:

    a. thymus gland

    b. ovary (plural—ovaries)

    c. testis (plural—testes)

*Remember to show this pre-lab to the SSL instructor **before entering** the lab. It will be marked off in the SSL book for grade credit towards your lab test.*

# 1

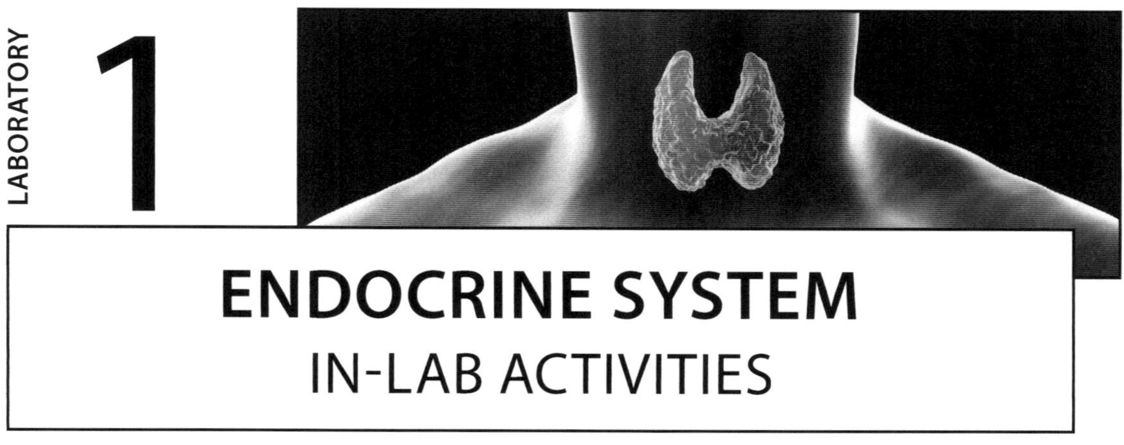

# ENDOCRINE SYSTEM
## IN-LAB ACTIVITIES

Name: _______________________________     Section: ___________     Date: _________

## LEARNING OBJECTIVES

1. Identify and characterize the two most important systems for maintaining homeostasis.

2. Identify and locate the glands which make up the endocrine system.

3. Detail the general functions of the hormones secreted by each endocrine gland.

4. Discuss the relationship between the hypothalamus and pituitary gland.

You will spend **2 hours or so** in lab at Forsyth Tech to complete the following activities. This amount of time allows you to complete the activities by using the torso model, using other models, and working with a lab partner.

## ACTIVITY 1: COMPARE AND CONTRAST THE NERVOUS AND ENDOCRINE SYSTEMS

*Use your text to answer the following questions. For question #2, use your lab models.*

1. Complete the summary table below that contrasts the nervous and endocrine systems.

|  | ENDOCRINE SYSTEM | NERVOUS SYSTEM |
|---|---|---|
| Primary Chemical Signal |  |  |
| Distance Traveled |  |  |
| Response Time |  |  |
| Response Duration |  |  |
| Distribution |  |  |

2. What structure connects the nervous system with the endocrine system?

 

3. Can a substance be classified as a hormone and a neurotransmitter? Defend your answer.

## ACTIVITY 2: IDENTIFY AND LOCATE ALL ENDOCRINE GLANDS

*You will need a full torso model (if available) to complete this exercise. Keep in mind that your lab test will be on the models from lab, pictures of those models, cat dissection, or other dissection.*

1. Obtain a full torso model (or cat dissection, if available) and locate the following glands: the ovaries, testes, thymus, hypothalamus, pancreas, pituitary, thyroid, parathyroid, adrenal, and pineal glands.

2. Label the figure below **and** under each gland you identified, list the hormones that are secreted from that particular gland. Be able to find these on the models in lab.

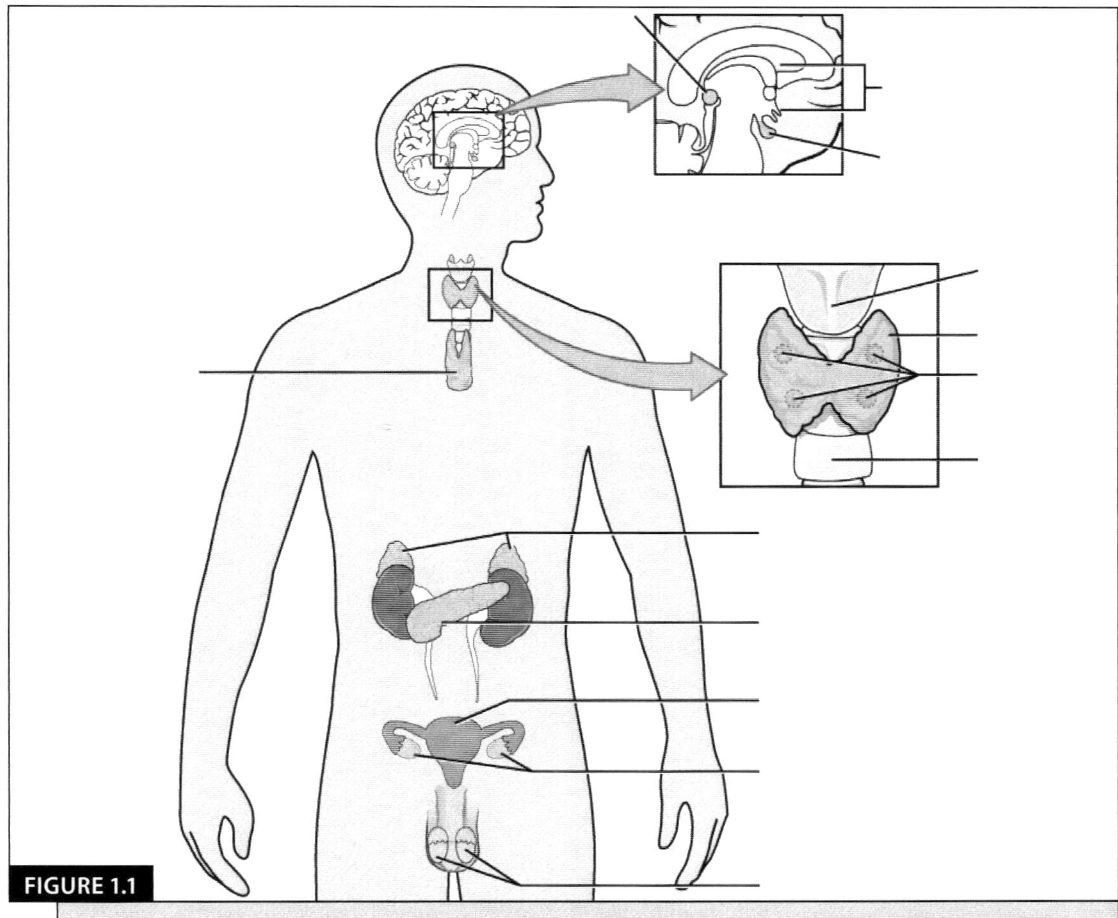

**FIGURE 1.1**

**Image of the human torso with the endocrine glands of the body marked for labeling**
OpenStax, An Overview of the Endocrine System. OpenStax CNX. Jun 18, 2013 http://cnx.org/contents/e250c2d0-97c5-4ec2-88fd-20207d1acdde@3

## ACTIVITY 3: EXAMINE THE RELATIONSHIP BETWEEN THE HYPOTHALAMUS AND PITUITARY GLANDS

*You will need a brain model, endocrine histology slides, and a compound light microscope (if available) to complete this exercise.*

1. Complete the summary table below.

| ENDOCRINE GLAND | HORMONE | HORMONE ACTION |
|---|---|---|
| Anterior Pituitary (Adenohypophysis) | | |
| | | |
| | | |
| | | |
| | | |
| | | |
| | | |
| Posterior Pituitary (Neurohypophysis) | | |
| | | |

2. Label the diagram below. *(Find these on the sagittal head model or torsos.)*

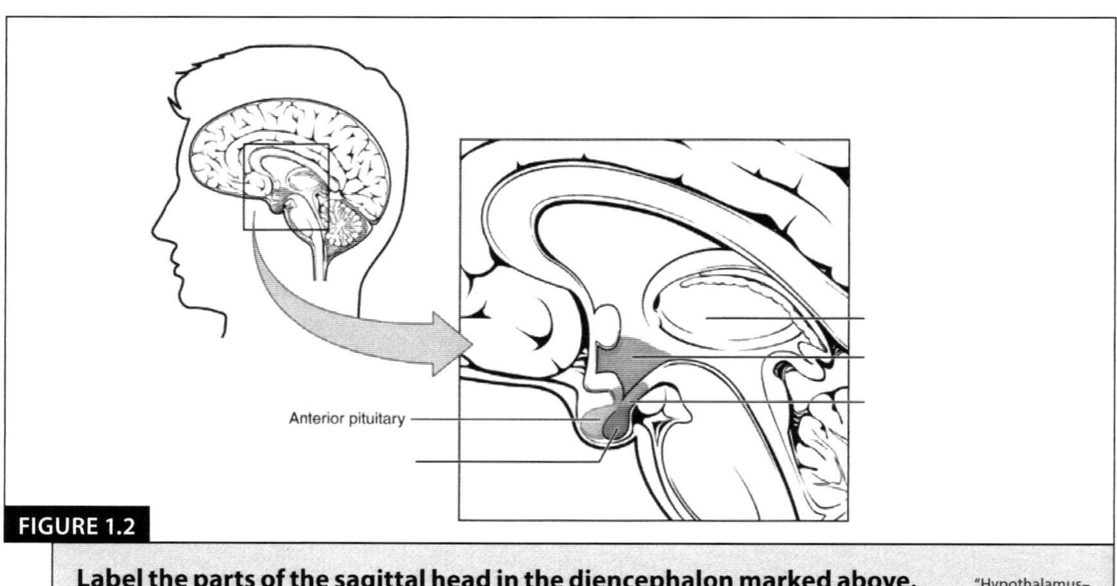


**FIGURE 1.2**

**Label the parts of the sagittal head in the diencephalon marked above.** "Hypothalamus–Pituitary Complex" by Phil Schatz. License: CC BY 4.0

3. Sketch in the circle below the histology of the adenohypophysis gland when viewed under the compound light microscope. (Note the darker stained tissue of this gland compared to the lighter stained tissue of the neurohypophysis.)

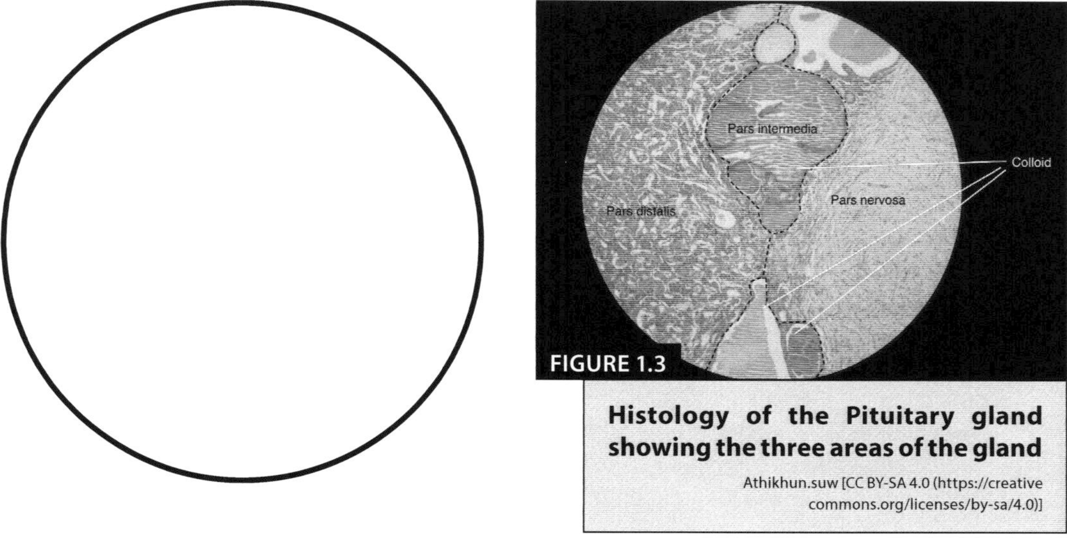


**FIGURE 1.3**

**Histology of the Pituitary gland showing the three areas of the gland**

Athikhun.suw [CC BY-SA 4.0 (https://creative commons.org/licenses/by-sa/4.0)]

4. Sketch in the circle below the histology of the neurohypophysis gland when viewed under the compound light microscope. (Note the lighter stained tissue of the neurohypophysis gland compared to the darker stained tissue of adenohypophysis cells.)

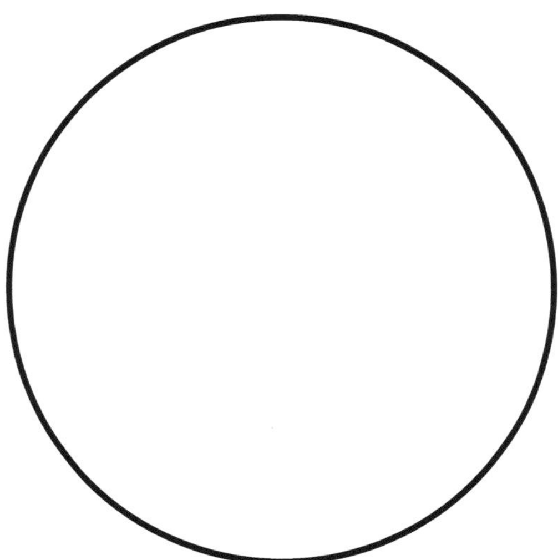

## ACTIVITY 4: DETAIL THE GENERAL FUNCTIONS OF THE HORMONES SECRETED BY EACH OF THE ENDOCRINE GLANDS

*You will need a full torso model, endocrine gland models, endocrine histology slides, and a compound light microscope (if available) to complete this exercise.*

1. Complete the summary table below.

| ENDOCRINE GLAND | HORMONE | HORMONE ACTION |
|---|---|---|
| Thyroid | | |
| Parathyroid | | |
| Adrenal | | |
| Pineal | | |
| Pancreas | | |
| Testes | | |
| Ovaries | | |

## ACTIVITY 5: IN THE CIRCLES PROVIDED, SKETCH THE HISTOLOGY OF EACH ENDOCRINE GLAND WHEN VIEWED UNDER THE COMPOUND LIGHT MICROSCOPE

*Notice cell shapes, layering, and other characteristics that distinguish each glandular tissue.*

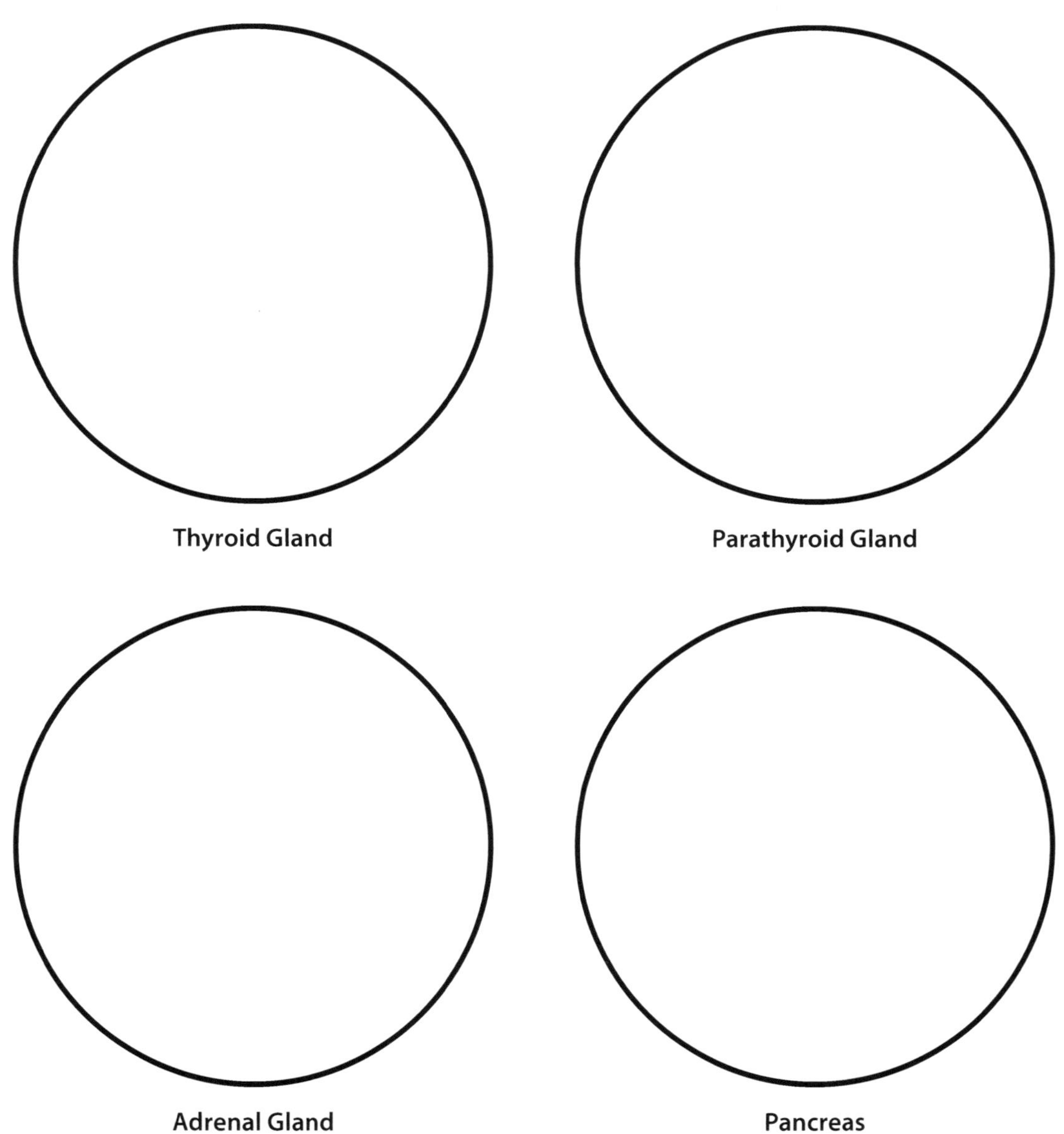

Thyroid Gland

Parathyroid Gland

Adrenal Gland

Pancreas

## EXTENSION QUESTIONS

*Complete these fully and in your own words.*

1. Define negative and positive feedback. In your answer, include an example of each type of feedback using hormones.

2. Name at least three organs with secondary endocrine function. In your answer, include the hormone that each organ secretes and the actions of each of those hormones.

3. Your patient is a 35-year-old female. She weighs 220 pounds and is 28 weeks pregnant. She failed her initial glucose screening and has returned for her 3 hour glucose tolerance test. She was instructed to fast and then drink a glucose solution. Her blood was taken before the test, one hour in, two hours in, and at the end of the 3 hours. Her results respectively are as follows: 80 mg/dL, 190 mg/dL, 140 mg/dL, and 85 mg/dL.

   - Does she have gestational diabetes? How do you know?

- What hormone level increased during the test?

- How did that hormone lower her blood sugar?

- What hormone is released by the pancreas during the times of fasting and several hours after eating?

- How does this hormone change the glucose levels in the blood stream?

4.  Your patient is a 27-year-old male weighing 110 pounds. His symptoms include sudden weight loss, sweating, and a heart rate of 105 bpm. There is a family history of Graves Disease. You send him for a radioactive iodine uptake test. His test results show a lower-than-normal uptake. You expected higher-than-normal uptake but still diagnose him with hyperthyroidism.

    - Explain why he could still have hyperthyroidism.

    - What are your recommendations for treatment?

**5.** Identify the following four glands by their histology.

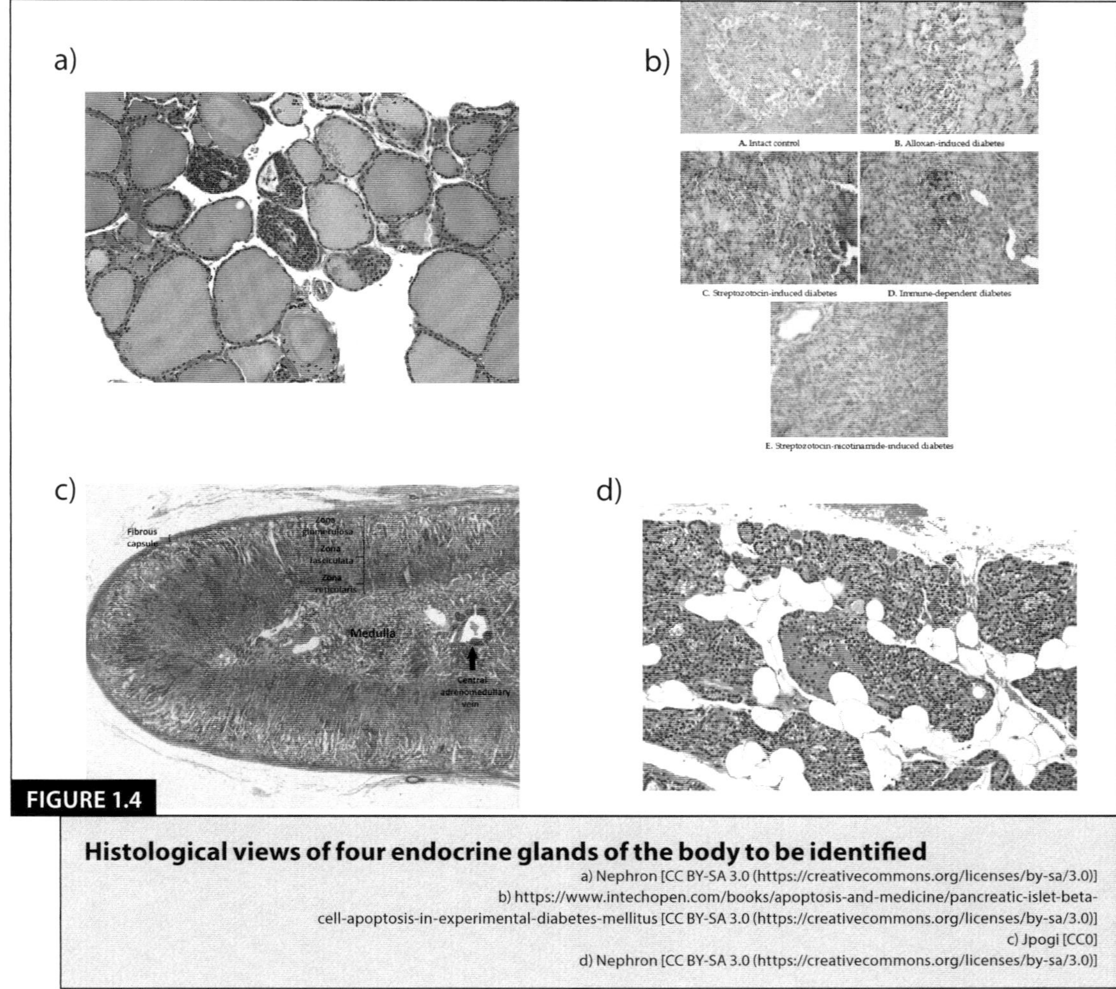



**FIGURE 1.4**

**Histological views of four endocrine glands of the body to be identified**

a) Nephron [CC BY-SA 3.0 (https://creativecommons.org/licenses/by-sa/3.0)]
b) https://www.intechopen.com/books/apoptosis-and-medicine/pancreatic-islet-beta-cell-apoptosis-in-experimental-diabetes-mellitus [CC BY-SA 3.0 (https://creativecommons.org/licenses/by-sa/3.0)]
c) Jpogi [CC0]
d) Nephron [CC BY-SA 3.0 (https://creativecommons.org/licenses/by-sa/3.0)]

*Be sure to get your completed work checked off by a member of the lab staff and then keep this hand out for your review.*

# 2

## BLOOD
### PRE-LAB

Name: _________________________   Section: __________   Date: ________

## LEARNING OBJECTIVES

1. Describe the physical characteristics of blood and its principle components.

2. Correctly perform a WBC differential using blood smear slides.

3. Explain how the different blood types are determined and perform blood typing with artificial blood.

4. Understand the importance of hematocrit and analyze test results.

*Checklist to complete* **before entering** *the science skills lab (SSL):*

☐ Actively read this packet of information.

☐ Complete the charts, tables or labeling and answer questions using your own words.

☐ Complete the electronic digital pre-lab quiz on Bb by Sunday.

☐ Review the attached anatomy list and take it to lab with you. Jot down key descriptive identifying words that aid in your lab test preparations.

## ACTIVITY 1: IDENTIFY THE PRINCIPLE COMPONENTS OF BLOOD

*Read the content in your text and the introductory paragraphs below. Then, complete Activity 1 in lab.*

Recall that **blood** is a connective tissue. In order to be considered a connective tissue the following must be present: cells, extracellular matrix surrounding the cells, and proteins either in fiber or dissolved form. The cellular elements—referred to as the **formed elements**—include **red blood cells (RBCs), white blood cells (WBCs),** and cell fragments called **platelets.** The extracellular matrix, called **plasma,** is fluid that perpetually suspends the formed elements and enables them to circulate throughout the body within the cardiovascular system.

The formed elements comprise roughly 37–54 percent of whole blood and are organized into 3 groups, each with its own function. The largest formed element group is the RBCs also known as **erythrocytes.** These cells have a 120 day life span, lack a nucleus or other mitotic equipment, have a unique shape, and iron atoms that allow them to efficiently perform their function of gas transportation. The second largest group is the WBCs, also known as **leukocytes.** There are five different types of leukocytes within this one group that function collectively for defense and immunity. The final group is cellular fragments called platelets or **thrombocytes.** These fragments do not contain a nucleus but contain granules that aide in their function of clotting.

From your reading above, fill in the table below with each component of blood and basic function.

| COMPONENT | SUBCOMPONENT | FUNCTION |
|---|---|---|
| Plasma | Water | |
| | Plasma Proteins | |
| | | |
| | | |
| Formed Elements | | |
| | | |
| | | |

When scientists first began to observe stained blood slides, it quickly became evident that leukocytes could be divided into two groups, according to whether their cytoplasm contained highly visible granules.

- **Granular leukocytes** contain abundant granules within the cytoplasm. They include neutrophils, eosinophils, and basophils.
- While granules are not totally lacking in **agranular leukocytes,** they are far fewer and less obvious. Agranular leukocytes include monocytes and lymphocytes.

# GRANULAR LEUKOCYTES

We will consider the granular leukocytes in order from most common to least common. All of these are produced in the red bone marrow and have a short lifespan of hours to days. They typically have a lobed nucleus and are classified according to which type of stain best highlights their granules.

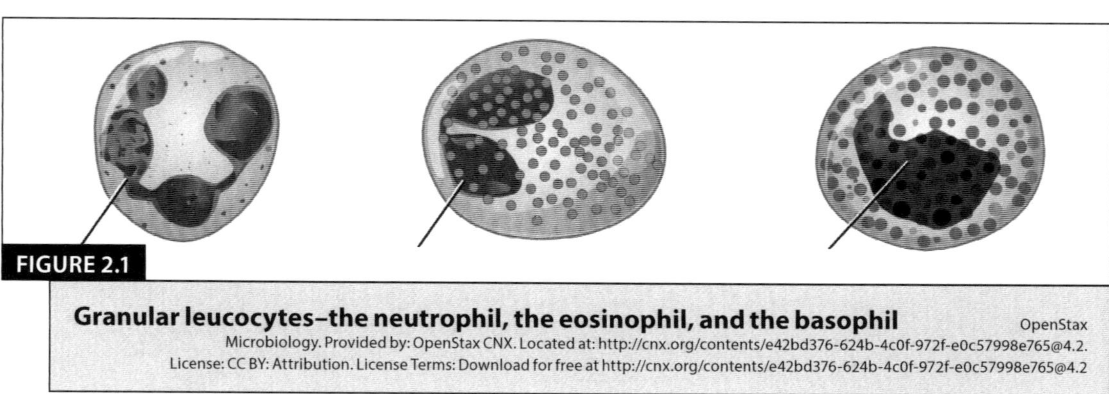

**FIGURE 2.1**

**Granular leucocytes–the neutrophil, the eosinophil, and the basophil**  OpenStax
Microbiology. Provided by: OpenStax CNX. Located at: http://cnx.org/contents/e42bd376-624b-4c0f-972f-e0c57998e765@4.2.
License: CC BY: Attribution. License Terms: Download for free at http://cnx.org/contents/e42bd376-624b-4c0f-972f-e0c57998e765@4.2

The most common of all the leukocytes, **neutrophils,** will normally comprise 50–70 percent of total leukocyte count. They are 10–12 µm in diameter and have granules that are numerous but quite fine and normally appear light lilac. The nucleus has a distinct lobed appearance and may have two to five lobes, the number increasing with the age of the cell. Young neutrophils are referred to as "bands," whereas the more aged cells with the lobed nucleus are referred to as "segs." Neutrophils are rapid responders to the site of infection and are efficient phagocytes with a preference for bacteria.

**Eosinophils** typically represent 2–4 percent of total leukocyte count. They are also 10–12 µm in diameter. The nucleus of the eosinophil will typically have two to three lobes and, if stained properly, the granules will have a distinct red to orange color. High counts of eosinophils are typical of patients experiencing allergies, parasitic worm infestations, and some autoimmune diseases.

**Basophils** are the least common leukocytes, typically comprising less than one percent of the total leukocyte count. They are slightly smaller than neutrophils and eosinophils at 8–10 µm in diameter. Basophils contain large granules that pick up a dark blue stain and are so common they may make it difficult to see the two-lobed nucleus. In general, basophils intensify the inflammatory response. The granules of basophils release histamines, which contribute to inflammation, and heparin, which opposes blood clotting. High counts of basophils are associated with allergies, parasitic infections, and hypothyroidism.

## AGRANULAR LEUKOCYTES

Agranular leukocytes contain smaller, less-visible granules in their cytoplasm than do granular leukocytes. The nucleus is simple in shape, sometimes with an indentation but without distinct lobes. There are two major types of agranulocytes: lymphocytes and monocytes.

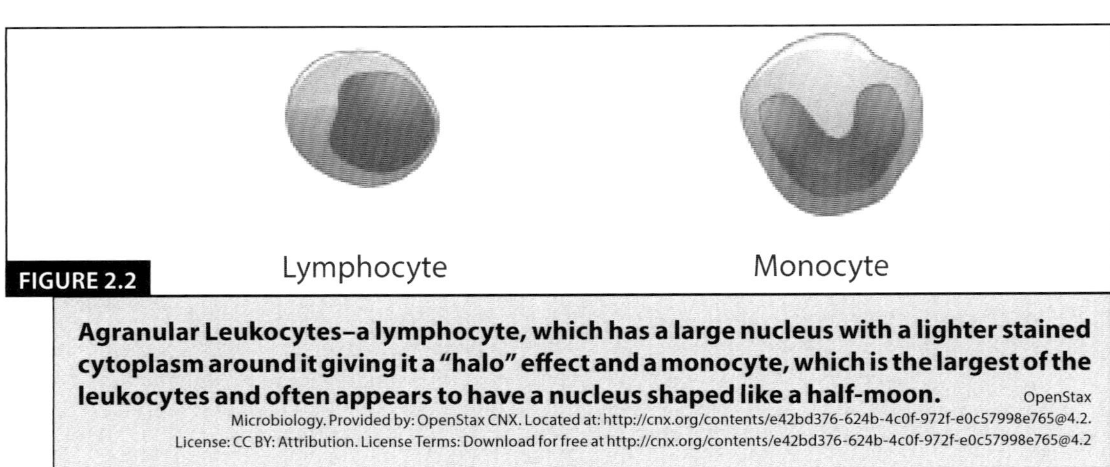


**FIGURE 2.2**

**Agranular Leukocytes–a lymphocyte, which has a large nucleus with a lighter stained cytoplasm around it giving it a "halo" effect and a monocyte, which is the largest of the leukocytes and often appears to have a nucleus shaped like a half-moon.** OpenStax Microbiology. Provided by: OpenStax CNX. Located at: http://cnx.org/contents/e42bd376-624b-4c0f-972f-e0c57998e765@4.2. License: CC BY: Attribution. License Terms: Download for free at http://cnx.org/contents/e42bd376-624b-4c0f-972f-e0c57998e765@4.2

**Lymphocytes** are the second most common type of leukocyte, accounting for about 20–30 percent of all leukocytes, and are essential for the acquired immune response. The size range of lymphocytes is quite extensive. Typically, the large cells are 10–14 µm and have a smaller nucleus-to-cytoplasm ratio and more granules. The smaller cells are typically 6–9 µm with a larger volume of nucleus to cytoplasm, creating a "halo" effect.

**Monocytes** originate from myeloid stem cells. They normally represent 2–8 percent of the total leukocyte count. They are typically easily recognized by their large size of 12–20 µm and indented or horseshoe-shaped nuclei. Macrophages are monocytes that have left the circulation and phagocytize debris, foreign pathogens, worn-out erythrocytes, and many other dead, worn out, or damaged cells.

Match the lymphocytes below with their correct functions.

1. Neutrophil ______ participate in acquired immune response as T cells and B cells

2. Eosinophil ______ inflammatory response from histamine release

3. Basophil ______ perform phagocytosis of bacteria

4. Monocyte ______ perform phagocytosis of debris, worn out cells, pathogens, etc.

5. Lymphocyte ______ numbers increase with autoimmune diseases

## ACTIVITY 2: EXPLAIN HOW THE DIFFERENT BLOOD TYPES ARE DETERMINED AND PERFORM BLOOD TYPING

*Read the content in your text and the introductory paragraphs below. Then, complete Activity 2 in lab.*

Although the **ABO blood group** name consists of three letters, ABO blood typing designates the presence or absence of just two antigen proteins, A and B. People whose erythrocytes have A antigen proteins on their erythrocyte membrane surfaces are designated blood type A, and those whose erythrocytes have B antigen proteins are blood type B. People can also have both A and B antigens on their erythrocytes, in which case they are blood type AB. People with neither A nor B antigens are designated blood type O.

ABO blood types are genetically determined. Individuals with type A blood—without any prior exposure to incompatible blood—have antibodies to the B antigen circulating in their blood plasma. These antibodies, referred to as anti-B antibodies, will cause agglutination and hemolysis if they ever encounter erythrocytes with B antigens. Similarly, an individual with type B blood has pre-formed anti-A antibodies. Individuals with type AB blood, which has both antigens, do not have antibodies to either of these. People with type O blood lack antigens A and B on their erythrocytes, but both anti-A and anti-B antibodies circulate in their blood plasma.

|  | Group A | Group B | Group AB | Group O |
|---|---|---|---|---|
| Red blood cell type | A | B | AB | O |
| Antibodies in plasma | Anti-B | Anti-A | None | Anti-A and Anti-B |
| Antigens in red blood cell | A antigen | B antigen | A and B antigens | None |

**FIGURE 2.3**

This chart shows the antibodies in plasma and antigens on the red blood cells that are present for each of the ABO blood types. [Public domain- InvictaHOG] https://commons.wikimedia.org/wiki/File:ABO_blood_type.svg

The **Rh blood group** is classified according to the presence or absence of a second erythrocyte antigen identified as Rh D. Those who have the Rh D antigen present on their erythrocytes—about 85 percent of Americans—are described as Rh positive (Rh+) and those who lack it are Rh negative (Rh–). Note that the Rh group is distinct from the ABO group, so any individual, no matter their ABO blood type, may have or lack this Rh antigen. When identifying a patient's blood type, the Rh group is designated by adding the word positive or negative to the ABO type. For example, A positive (A+) means ABO group A blood with the Rh antigen present, and AB negative (AB–) means ABO group AB blood without the Rh antigen.

Clinicians are able to determine a patient's blood type quickly and easily using commercially prepared antibodies. An unknown blood sample is allocated into separate wells. Into one well a small amount of anti-A antibody is added and to another, a small amount of anti-B antibody. If the antigen is present, the antibodies will cause visible agglutination of the cells. The blood should also be tested for Rh antibodies by adding an anti-D antibody to the last well.

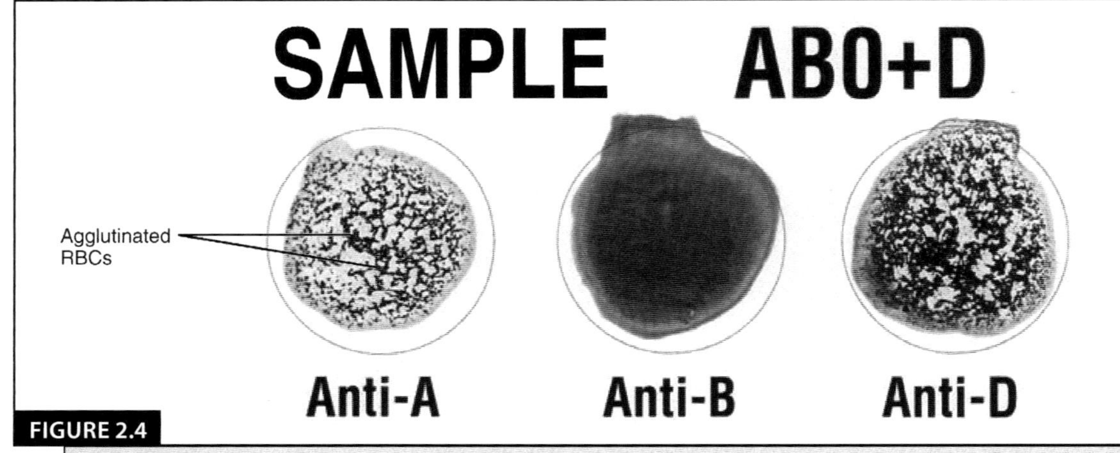


**FIGURE 2.4**

**This sample of a commercially produced "bedside" card enables quick typing of both a recipient's and donor's blood before transfusion.** The card contains three reaction sites or wells. One is coated with an anti-A antibody, one with an anti-B antibody, and one with an anti-D antibody (tests for the presence of Rh factor D). Mixing a drop of blood and saline into each well enables the blood to interact with a preparation of type-specific antibodies, also called anti-seras. Agglutination of RBCs in a given site indicates a positive identification of the blood antigens, in this case A and Rh antigens for blood type A+. For the purpose of transfusion, the donor's and recipient's blood types must match. https://opentextbc.ca/anatomyandphysiology/chapter/18-6-blood-typing/ [CC BY-SA 4.0 (https://creativecommons.org/licenses/by-sa/4.0)]

Complete the summary table below by listing the compatible blood types of each blood type on the left. Also, recall the anti-bodies that will react against each type.

| BLOOD TYPE | COMPATIBLE BLOOD TYPES |
|---|---|
| A positive | |
| A negative | |
| B positive | |
| B negative | |
| AB positive | |
| AB negative | |
| O positive | |
| O negative | |

## ACTIVITY 3: UNDERSTAND THE IMPORTANCE OF HEMATOCRIT AND ANALYZE TEST RESULTS

*Read the content in your text and the introductory paragraphs below. Then, complete Activity 3 in lab.*

A **hematocrit** measures the percentage of RBCs, clinically known as erythrocytes, in a blood sample. It is performed by spinning the blood sample in a specialized centrifuge, a process that causes the heavier elements suspended within the blood sample to separate from the lightweight, liquid plasma. Because the heaviest elements in blood are the erythrocytes, these settle at the very bottom of the hematocrit tube. Located above the erythrocytes is a pale, thin layer composed of the remaining formed elements of blood. These are the WBCs, clinically known as leukocytes, and the platelets, cell fragments also called thrombocytes. This layer is referred to as the **buffy coat** because of its color; it normally constitutes less than 1 percent of a blood sample. Above the buffy coat is the blood plasma, normally a pale, straw colored fluid, which constitutes the remainder of the sample.

The volume of erythrocytes after centrifugation is also commonly referred to as **packed cell volume (PCV).** In normal blood, about 45 percent of a sample is erythrocytes. The hematocrit of any one sample can vary significantly; about 36–50 percent, according to gender and other factors. Normal hematocrit values for females range from 37 to 47 with a mean value of 41; for males, hematocrit ranges from 42 to 52 with a mean of 47. The percentage of other formed elements, the WBCs and platelets, is extremely small so it is not normally considered with the hematocrit. So, the mean plasma percentage is the percent of blood that is not erythrocytes; for females, it is approximately 59 (or 100 minus 41), and for males, it is approximately 53 (or 100 minus 47).

**Preview** these terms in the reading above and in the open source text for the course before attending your lab. Once you have completed the Pre-Lab quiz online, you will be able to open the handout for your lab activities.

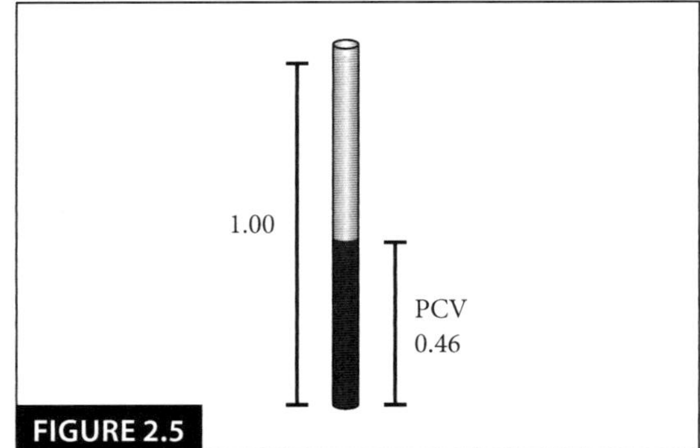


**FIGURE 2.5**

**This diagram represents a hematocrit capillary tube.** Whole blood was collected and centrifuged in this capillary tube. Next, the packed cell volume (PCV) is measured in millimeters. Hematocrit is calculated by dividing the length of the PCV by the length of the whole column (buffy coat and PCV); then, multiply that answer by 100.

## ANATOMY LIST

1.  Identify the principle components of blood:
    a.  plasma
    b.  serum
    c.  formed elements (RBCs, WBCs, and platelets)
2.  Perform a WBC differential, identify in a blood sample, and know the functions of the following cells:
    a.  erythrocytes
    b.  thromobcytes
    c.  granular leukocytes
        i.  neutrophil
        ii.  eosinophil
        iii.  basophil

    **d.** agranular leukocytes

        **i.** lymphocyte

        **ii.** monocyte

**3.** Identify the following blood types from scenarios or blood typing experiments:

    **a.** A positive and negative

    **b.** B positive and negative

    **c.** AB positive and negative

    **d.** O positive and negative

**4.** Identify the following portions of a hematocrit tube, calculate hematocrit values, and interpret results:

    **a.** plasma

    **b.** buffy coat

    **c.** PCV

    **d.** anemia

    **e.** polycythemia

*Remember to show this pre-lab to the SSL instructor* **before entering** *the lab. It will be marked off in the SSL book for grade credit towards your lab test.*

# 2

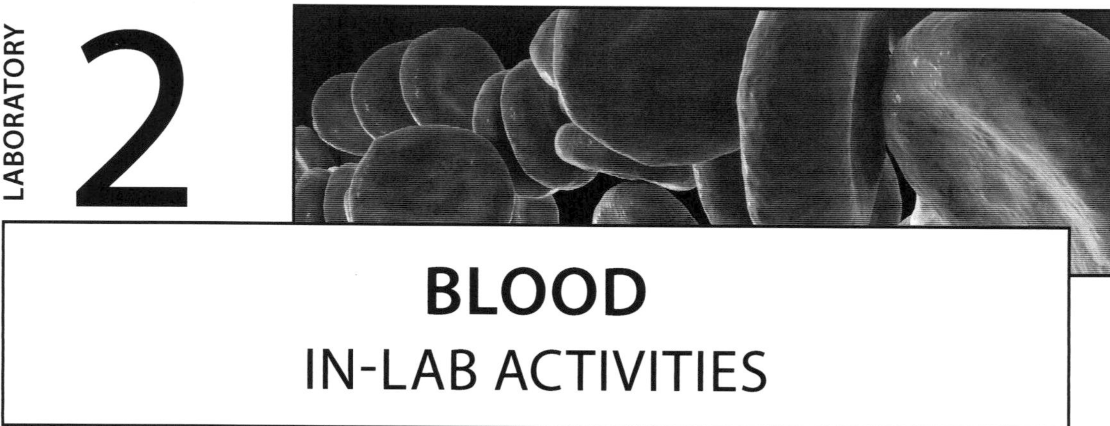

# BLOOD
## IN-LAB ACTIVITIES

Name: _______________________    Section: __________    Date: _________

## LEARNING OBJECTIVES

1. Describe the physical characteristics of blood and its principle components.

2. Correctly perform a WBC differential using blood smear slides.

3. Explain how the different blood types are determined and perform blood typing with artificial blood.

4. Understand the importance of hematocrit and analyze test results.

## ACTIVITY 1: IDENTIFY THE PRINCIPLE COMPONENTS OF BLOOD

*You will need blood smear slides and a compound light microscope (if available).*

1. Label the diagram of whole blood below–this sample was centrifuged. Recall from the Pre-Lab that the "buffy coat" is above the PCV.

2. Obtain a blood histology slide. Examine the slide with the compound light microscope. Scan through the same and locate each of the formed elements in the table below. Then, complete the table, below drawing what the formed element looked like under the microscope and giving the function for each leukocyte.

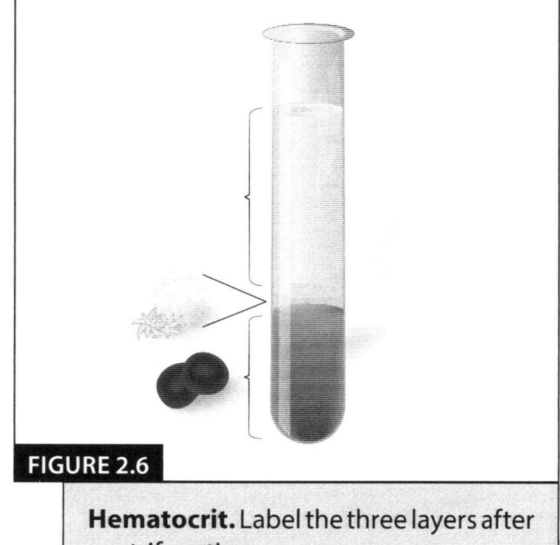

**FIGURE 2.6**

**Hematocrit.** Label the three layers after centrifugation.



ANATOMY AND PHYSIOLOGY II **33**

3. Complete the table.

| LEUKOCYTE | DRAWING | FUNCTION |
|---|---|---|
| Neutrophil | | |
| Lymphocyte | | |
| Monocyte | | |
| Eosinophil | | |
| Basophil | | |

4. Perform a WBC differential. Complete the chart below and answer the additional questions.

   a. Obtain two blood smear slides–one normal and the other abnormal (You may choose either the sickle cell anemia or mononucleosis positive sample. These are available in the lab or in your the lab box.)

   b. Bring the normal sample into focus under the scanning objective lens using a compound light microscope.

**c.** Next, move up to high power and bring into focus (Remember to pass through low power first.)

**d.** Then, starting at the top left hand corner of your sample, move horizontally to the right. As you slowly move to the right, identify all leukocytes that enter the field of view and mark them in the table below with a tally mark i.e., if a neutrophil enters your field of view, put a tally mark in the neutrophil column.

**e.** When you reach the end of the sample, move vertically down further into the sample.

**f.** Begin scanning back to the left and continue to mark each leukocyte with a tally mark (see below for the scanning method).

**g.** You will continue this process horizontally scanning and dropping vertically until just **100 leukocytes** have been tallied.

**h.** Complete the table for the normal sample following the instructions in the remaining columns.

**i.** Repeat the entire process using your abnormal sample.

**j.** Complete both tables and answer the discussion questions.

**k.** Scanning Method—follow the arrows on the slide and sample below.

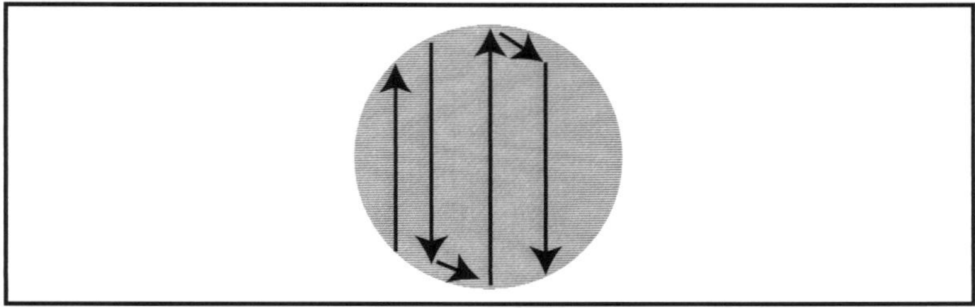

**5.** Differential Data Collection Tables

| NORMAL BLOOD SAMPLE | | | |
|---|---|---|---|
| **TYPE OF WBC** | **TALLY MARKS** | **% OF TOTAL (TALLY / 100 = %)** | **NORMAL % (LOOK THIS VALUE UP IN YOUR TEXT)** |
| Neutrophil | | | |
| Lymphocyte | | | |
| Monocyte | | | |
| Eosinophil | | | |
| Basophil | | | |

| ABNORMAL BLOOD SAMPLE | | | |
|---|---|---|---|
| **TYPE OF WBC** | **TALLY MARKS** | **% OF TOTAL (TALLY / 100 = %)** | **NORMAL % (SAME VALUES AS PREVIOUS TABLE)** |
| Neutrophil | | | |
| Lymphocyte | | | |
| Monocyte | | | |
| Eosinophil | | | |
| Basophil | | | |

## DISCUSSION QUESTIONS

1.  Which abnormal sample did you choose?

    _______________________________________________________________

2.  What values in this abnormal sample deviated from the normal values?

    _______________________________________________________________

    _______________________________________________________________

    _______________________________________________________________

    _______________________________________________________________

3.  Why were those particular leukocytes higher or fewer in concentration because of that particular illness? Be specific and thorough in your answer.

    _______________________________________________________________

    _______________________________________________________________

    _______________________________________________________________

    _______________________________________________________________

## ACTIVITY 2: EXPLAIN HOW THE DIFFERENT BLOOD TYPES ARE DETERMINED AND PERFORM BLOOD TYPING

*You will need a Carolina Artificial Blood Typing kit (if available).*

1. Perform a blood typing experiment on four different samples. Follow the Carolina Artificial Blood Typing experiment directions. Record your observations below and answer the discussion questions.

2. Data Collection Tables–indicate a positive or negative reaction to the added antibodies with a "+" or "−" sign in the appropriate columns. Then determine the blood type of your sample based on your observations.

| SAMPLE # | ANTI–A | ANTI–B | ANTI–RH (D) | IDENTIFIED BLOOD TYPE |
|---|---|---|---|---|
| 1 | | | | |
| 2 | | | | |
| 3 | | | | |
| 4 | | | | |

## DISCUSSION QUESTIONS

1. A patient with type A positive blood enters the emergency room and needs whole blood. What compatible blood types can you safely give them?

2. A patient with AB negative needs a blood transfusion. Her husband, who is A negative, and her brother, who is B positive, have volunteered as donors. Which donation can she safely accept and why?

3. A patient with type O positive blood enters the emergency room and needs whole blood. What compatible blood types can you safely give them?

4. Which blood type is considered the universal donor and why?

5. Which blood type is considered the universal recipient and why?

## ACTIVITY 3: UNDERSTAND THE IMPORTANCE OF HEMATOCRIT AND ANALYZE TEST RESULTS

*You will need capillary tubes that contain a blood sample and a ruler (if available).*

1. For the diagrams below, identify the condition that the patient is suffering from and give a brief explanation as to why you chose that condition. (A normal sample for reference is on the far left.)

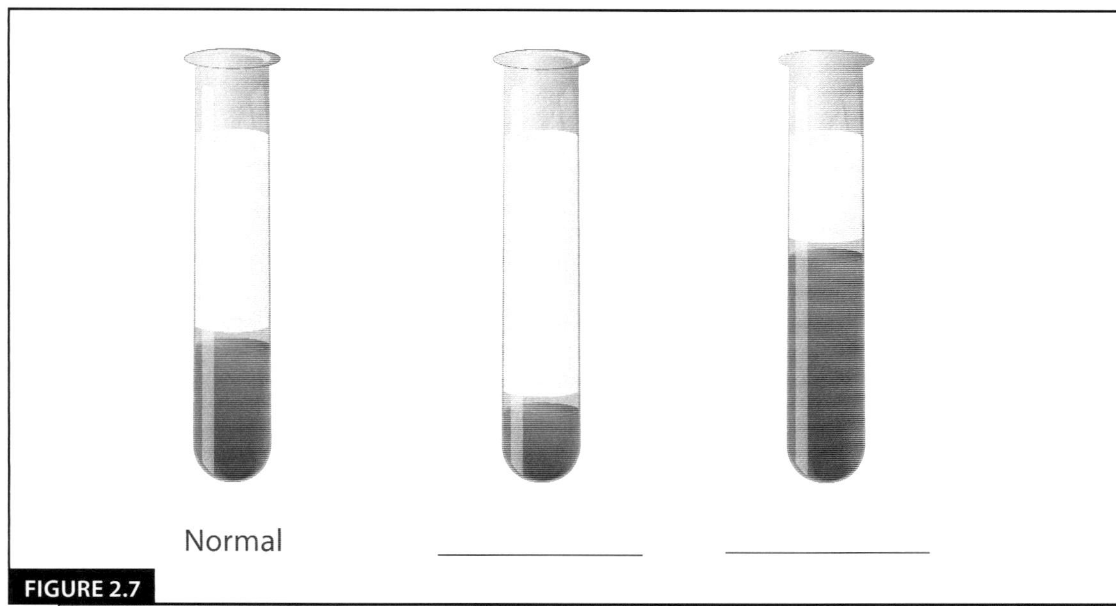


**FIGURE 2.7**

**Hematocrit.** This is the ratio of the volume of red cells to the volume of whole blood. A **low hematocrit** means the percentage of red blood cells is below the lower limits of normal, and is also called **anemia.** A **high hematocrit** is called **polycythemia.**

Explanations: _______________________________________________

_______________________________________________

_______________________________________________

_______________________________________________

_______________________________________________

_______________________________________________

2. Calculate hematocrit values for a blood sample. Given the hematocrit capillary tube below …

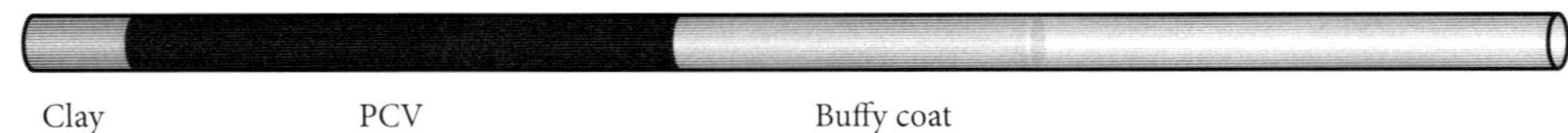


a. Using a ruler, measure the length of the RBCs in mm.

b. Then measure the length of the whole column (from air to clay) in mm.

c. Using the formula below, calculate the hematocrit value of the sample.

d. (Step A / Step B) × 100 = Hematocrit value

e. Place your work below.

3. Do you think the sample was taken from a man or a woman, or are you unable to tell? Why?

## EXTENSION QUESTIONS

*Complete these fully and in your own words.*

1.  When you are suffering from allergies, you take an anti-histamine such as Benadryl. Which leukocyte actions are you attempting to reverse by using this medication? Why?

2.  A patient's hematocrit is 42 percent. Approximately what percentage of the patient's blood is plasma?

3.  Sickle cell is the most common inherited disorder in the US despite its recessive genetic nature. This is, in part, due to the fact that sickle cell has an evolutionary plus. The sufferer inherits a type of resistance to malaria. Why would sickle cell help limit malaria's action in the human body?

4.  In a heparinized tube, how is coagulation time affected? Explain your answer.

*Be sure to get your completed work checked off by a member of the lab staff and then keep this hand out for your review.*

# 3

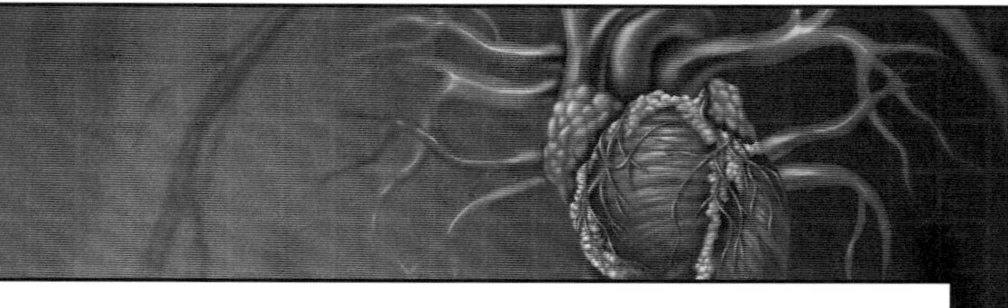

# THE HEART
## PRE-LAB

Name: _________________________  Section: __________  Date: _________

*The activities in this lab are associated with the lecture materials in the unit on the heart.*

## LEARNING OBJECTIVES

1. Identify major heart structures and know their functions, as well as the heart wall layers.
2. Outline the flow of blood through the heart.
3. Become familiar with fetal circulation and heart maturation.
4. Explain and identify vessels involved in coronary circulation.
5. Describe microscopic structures of the heart muscles.

*Checklist to complete* **before entering** *the science skills lab (SSL):*

☐ Actively read this packet of information.

☐ Complete the charts, tables or labeling and answer questions using your own words.

☐ Complete the electronic digital pre-lab quiz on Bb by Sunday.

☐ Review the attached anatomy list and take it to lab with you. Jot down key descriptive identifying words that aid in your lab test preparations.

☐ Watch these two videos at home: http://tinyurl.com/SSL—Blood Flow and http://tinyurl.com/SSL—Heart

# ACTIVITY 1: IDENTIFY MAJOR HEART STRUCTURES AND KNOW THEIR FUNCTIONS

*Read the content in your text and the introductory paragraphs below. Then, complete Activity 1 in lab.*

The human heart is located within the thoracic cavity, medially between the lungs in the space known as the mediastinum. It is about the size of a fist, is broad at the top, and tapers toward the base. The **base** of the heart is located at the level of the third costal cartilage. The inferior tip of the heart, the **apex,** lies just to the left of the sternum between the junction of the fourth and fifth ribs near their articulation with the costal cartilages. The apex deviates slightly to the left.

Within the mediastinum, the heart is separated from the other mediastinal structures by a tough membrane known as the pericardium, or pericardial sac, and sits in its own space called the **pericardial cavity.** The pericardium, which literally translates as "around the heart," consists of two distinct sublayers: the sturdy outer fibrous pericardium and the inner serous pericardium. The fibrous pericardium is made of tough, dense connective tissue that protects the heart and maintains its position in the thorax. The more delicate serous pericardium consists of two layers: the parietal pericardium, which is fused to the fibrous pericardium, and an inner visceral pericardium, or **epicardium,** which is fused to the heart and is part of the heart wall. The pericardial cavity, filled with lubricating serous fluid, lies between the **epicardium** and the visceral pericardium.

Label the layers on the figure below.

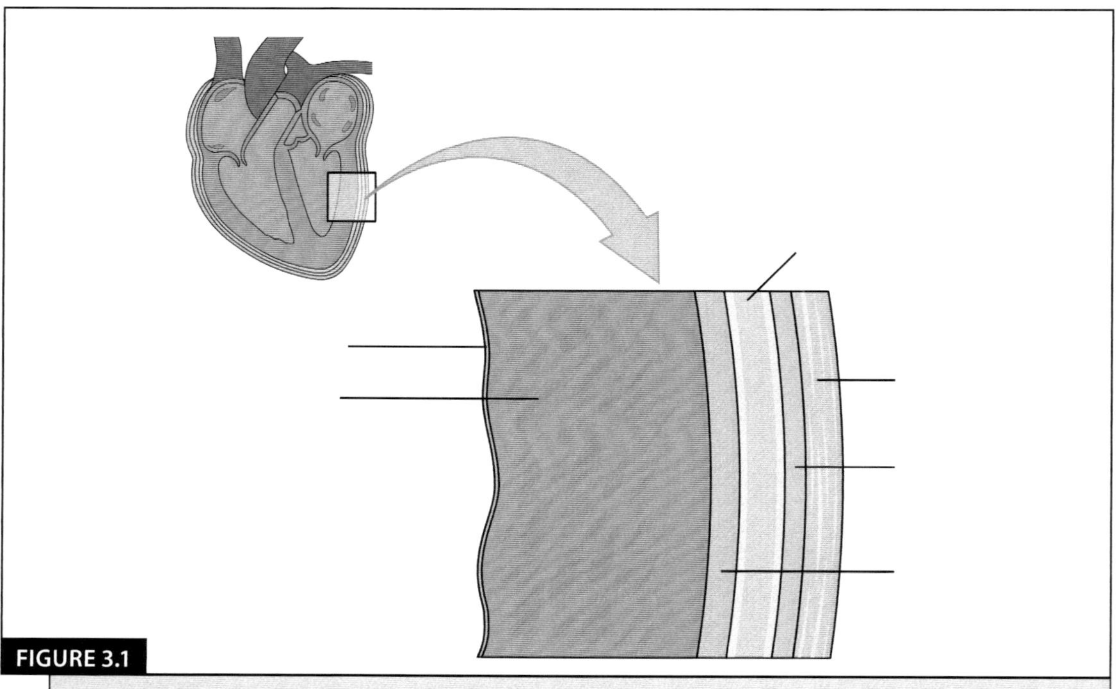

**FIGURE 3.1**

**Heart Wall.** Longitudinal section of the heart wall showing the epicardium, pericardium, and the myocardium to be labeled.   OpenStax College [CC BY 3.0 (https://creativecommons.org/licenses/by/3.0)]

Inside the pericardium, the surface features of the heart are visible, including the four chambers. There is a superficial leaf-like extension of the atria near the superior surface of the heart, one on each side, called an **auricle**—a name that means "ear like"—because its shape resembles the external ear of a human. Auricles are relatively thin-walled structures that can fill with blood and empty into the atria or upper chambers of the heart. You may also hear them referred to as atrial appendages.

Also prominent is a series of fat-filled grooves, each of which is known as a **sulcus** (plural = sulci), along the superior surfaces of the heart. Major coronary blood vessels are located in these sulci. The deep **coronary sulcus** is located between the atria and ventricles. Located between the left and right ventricles are two additional sulci that are not as deep as the coronary sulcus. The **anterior interventricular sulcus** is visible on the anterior surface of the heart, whereas the **posterior interventricular sulcus** is visible on the posterior surface of the heart.

Label the structure on the figure below.

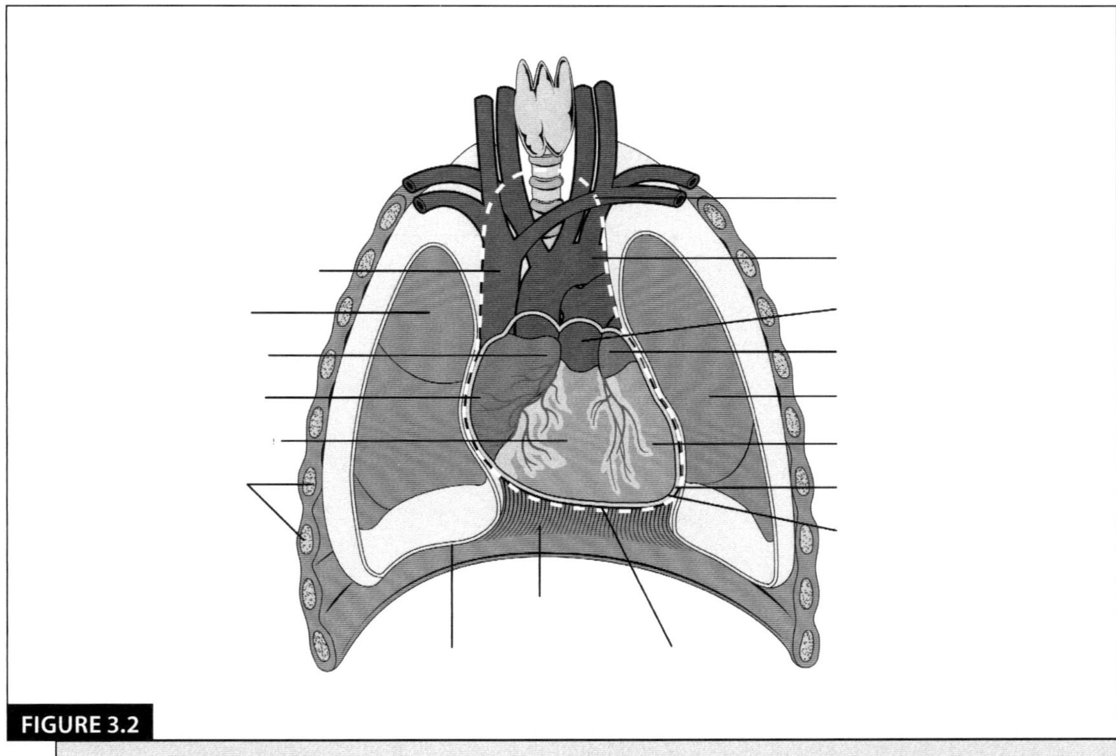

**FIGURE 3.2**

**The thoracic cavity including the lungs, blood vessels, and heart.** The heart is located within the thoracic cavity, medially between the lungs in the mediastinum. It is about the size of a fist, is broad at the top, and tapers toward the base. Heart Anatomy [https://cnx.org/contents/ FPtK1zmh@8.108:Y5T_wVSC@4/Heart-Anatomy] by OpenStax, (CC BY 4.0)

Label the parts of the heart and associated areas of the thoracic cavity.

# ACTIVITY 2: OUTLINE THE FLOW OF BLOOD THROUGH THE HEART

*Read the content in your text and the introductory paragraphs below. Then, complete Activity 2 in lab.*

The circulatory system is an organ system that passes nutrients, gases, hormones, and cells to and from tissues in the body to help fight diseases and stabilize body temperature and pH to maintain homeostasis.

There are two distinct, but linked, circuits in the human circulation called the pulmonary and systemic circuits. The **pulmonary circuit** is the portion of the cardiovascular system which transports oxygen-depleted blood away from the heart and to the lungs, and returns oxygenated blood back to the heart. The **systemic circuit** transports oxygenated blood to virtually all of the tissues of the body and returns relatively deoxygenated blood and carbon dioxide to the heart to be sent back to the pulmonary circulation.

Oxygen-deprived blood comes from tissues where it travels to the vena cava, right atrium, right ventricle, and into the pulmonary semilunar valve into the pulmonary arteries, which go to the lungs. In the lungs, blood gives up carbon dioxide and takes on a fresh supply of oxygen. Pulmonary veins return the now oxygen-rich blood to the heart, where it enters the left atrium before flowing through the mitral valve into the left ventricle where it is pumped out via the aorta and on to the rest of the body. Two openings lead to the right and left coronary arteries, which supply blood to the heart itself. From the heart, blood flows through the large aorta to the muscular arteries, which branch and narrow into the arterioles and then branch further still into the capillaries which are small and thin enough to perform gas exchange at various body tissues. On the way from the tissues to the heart, capillaries join and widen to become venules, and then widen more to become veins, which return blood to the heart. The heart is asymmetrical, with the left side being larger and stronger than the right side, so it can pump blood to the entire body.

The right atrium receives deoxygenated blood from the systemic circulation through the major veins: the superior vena cava, which drains blood from the head and from the veins that come from the arms, as well as the inferior vena cava, which drains blood from the veins that come from the lower organs and the legs. This deoxygenated blood then passes to the right ventricle through the tricuspid valve, which prevents the backflow of blood. After it is filled, the right ventricle contracts, pumping the blood to the lungs for reoxygenation. The left atrium receives the oxygen-rich blood from the lungs. This blood passes through the bicuspid valve to the left ventricle where the blood is pumped into the aorta. The aorta is the major artery of the body, taking oxygenated blood to the organs and muscles of the body. This pattern of pumping is referred to as double circulation and is found in all mammals.

Fill in other arrows for where blood goes from the right side of the heart to the left side and out to the body.

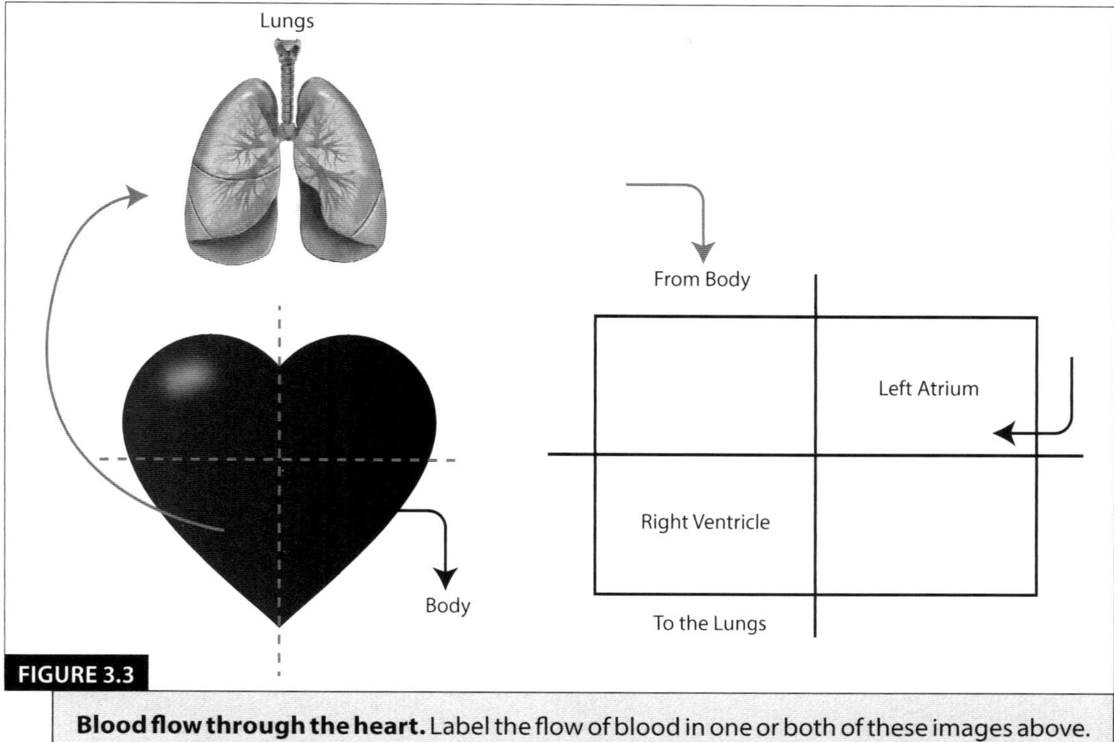


**FIGURE 3.3**

**Blood flow through the heart.** Label the flow of blood in one or both of these images above.

## ACTIVITY 3: BECOME FAMILIAR WITH FETAL CIRCULATION AND HEART MATURATION

*Read the content in your text and the introductory paragraphs below. Then, complete Activity 3 in lab.*

You will study the fetal heart structures as well. Look at the fetal circulation in the image below and find the three additional structures of the fetal heart. One such additional channel or opening allows for passage of blood to bypass the lungs for oxygenation: the foramen ovale. The **fossa ovalis** is the remnant of a thin fibrous sheet that covered the **foramen ovale** (oval opening) during fetal development.

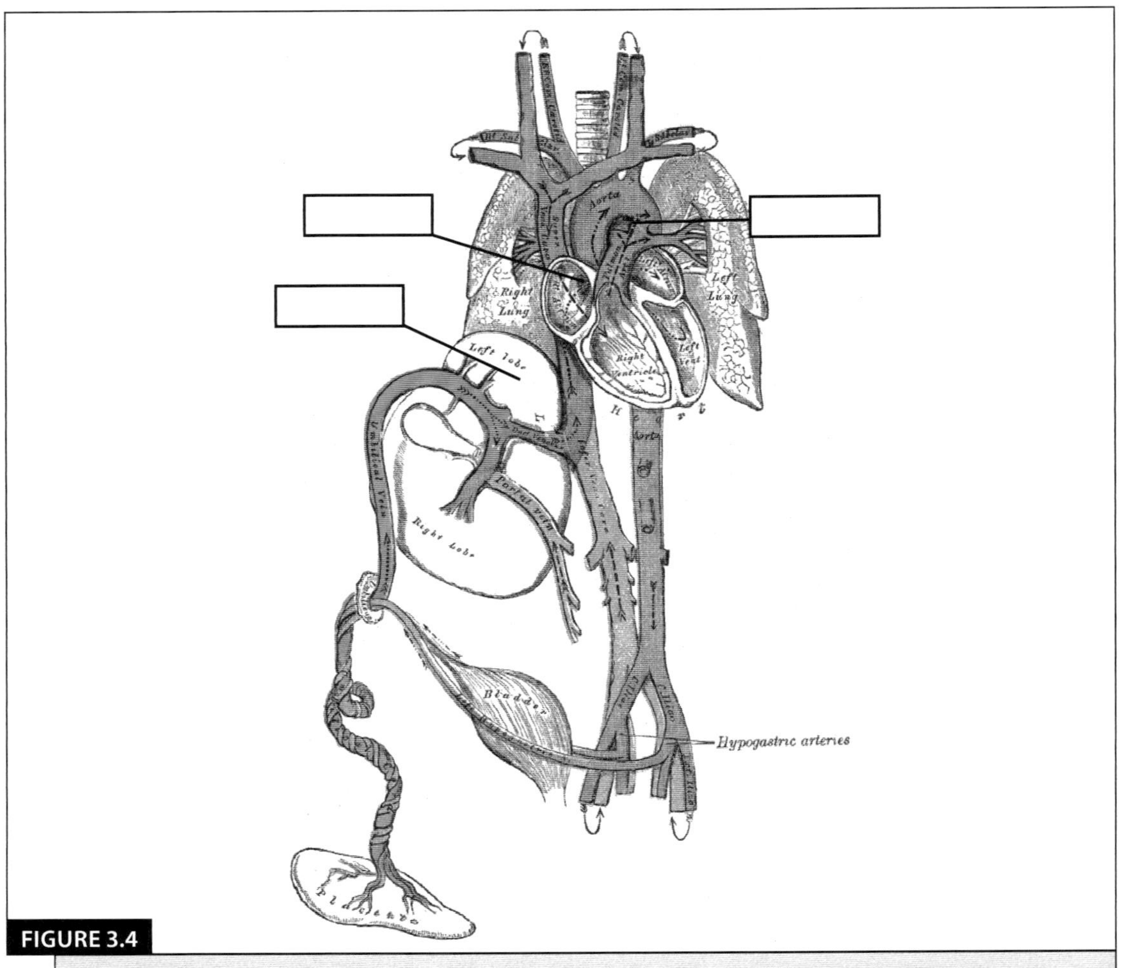



**FIGURE 3.4**

**Fetal circulation of heart showing both the ductus arteriosus and the foramen ovale in addition to the ductus venous to the placenta, which the student is to label.**

[Public domain, https://commons.wikimedia.org/wiki/File:Fetal_circulation.png.]

- Where is the foramen ovale found in the fetal heart? In lab, you can find the ovale or the fossa ovalis in the heart models or the heart dissection.

- What are the other two structures/ducts that are found associated with the fetal heart?

- Label the three boxes above.

## ACTIVITY 4: EXPLAIN AND IDENTIFY VESSELS INVOLVED IN CORONARY CIRCULATION

*Read the content in your text and the introductory paragraphs below. Then, complete Activity 4 in lab.*

**Coronary arteries** supply blood to the myocardium and other components of the heart. The left coronary artery distributes blood to the left side of the heart, the left atrium and ventricle, and the interventricular septum. The **circumflex artery** arises from the left coronary artery and follows the coronary sulcus to the left. Eventually, it will fuse with the small branches of the right coronary artery. The larger **anterior interventricular artery,** also known as the left anterior descending artery (LAD), is the second major branch arising from the left coronary artery. It follows the anterior interventricular sulcus around the pulmonary trunk.

The right coronary artery proceeds along the coronary sulcus and distributes blood to the right atrium, portions of both ventricles, and the heart conduction system. Normally, one or more marginal arteries arise from the right coronary artery inferior to the right atrium. The **marginal arteries** supply blood to the superficial portions of the right ventricle. On the posterior surface of the heart, the right coronary artery gives rise to the **posterior interventricular artery,** also known as the posterior descending artery. It runs along the posterior portion of the interventricular sulcus toward the apex of the heart, giving rise to branches that supply the interventricular septum and portions of both ventricles.

**Coronary veins** drain the heart and generally parallel the large surface arteries. The **great cardiac vein** can be seen initially on the surface of the heart following the anterior interventricular sulcus, but it eventually flows along the coronary sulcus into the coronary sinus on the posterior surface. (In order to recall this vein's location, think "it's great to be in front!") The **middle cardiac vein** parallels and drains the areas supplied by the posterior interventricular artery. The **small cardiac vein** parallels the right coronary artery and drains the blood from the posterior surfaces of the right atrium and ventricle. The **coronary sinus** is a large, thin-walled vein on the posterior surface of the heart lying within the atrioventricular sulcus and emptying directly into the right atrium.

- Why does the heart need to have its own circulation of blood vessels?

# ACTIVITY 5: DESCRIBE MICROSCOPIC STRUCTURES OF THE HEART MUSCLES

*Read the content in your text and the introductory paragraphs below. Then, complete Activity 5 in lab.*

Recall that cardiac muscle shares a few characteristics with both skeletal muscle and smooth muscle, but it has some unique properties of its own. Not the least of these exceptional properties is its ability to initiate an electrical potential at a fixed rate that spreads rapidly from cell to cell to trigger the contractile mechanism. This property is known as **autorhythmicity.** Even though cardiac muscle has autorhythmicity, heart rate is modulated by the endocrine and nervous systems.

Compared to the giant cylinders of skeletal muscle, cardiac muscle cells, or cardiomyocytes, are considerably shorter with much smaller diameters. Cardiac muscle also demonstrates striations, the alternating pattern of dark A bands and light I bands attributed to the precise arrangement of the myofilaments and fibrils that are organized in sarcomeres along the length of the cell. These contractile elements are virtually identical to skeletal muscle. Typically, cardiomyocytes have a single, central nucleus, but two or more nuclei may be found in some cells. Cardiac muscle cells branch freely. A junction between two adjoining cells is marked by a critical structure called an **intercalated disc,** which helps support the synchronized contraction of the muscle. They consist of desmosomes, specialized linking proteoglycans, tight junctions, and large numbers of gap junctions that allow the passage of ions between the cells and help to synchronize the contraction.

- What are the structures of the intercalated discs, and which part is most important for communication between cardiac cells?

*Checklist to complete* **before entering** *the science skills lab (SSL):*

- ☐ Actively read this packet of information.
- ☐ Complete the charts, tables or labeling and answer questions using your own words.
- ☐ Complete the electronic digital pre-lab quiz on Bb by Sunday.
- ☐ Review the attached anatomy list and take it to lab with you. Jot down key descriptive identifying words that aid in your lab test preparations.
- ☐ Watch these two videos at home: http://tinyurl.com/SSL—Blood Flow and http://tinyurl.com/SSL—Heart

## ACTIVITY 6: IDENTIFY ALL ANATOMICAL STRUCTURES OF THE HEART LISTED BELOW WHEN IN THE LAB (OR PERFORMING A DISSECTION).

### ANATOMY LIST

1. Identify the following **microscopic structures** on a slide of cardiac muscle tissue:
   a. cardiac muscle fibers (cardiocytes)

      **Note** the branching arrangement of the cardiocytes.
   b. intercalated discs
2. Identify the following **cardiac structures** on models and diagrams:
   a. pericardium
      i. parietal layer (fibrous and serous layers)
      ii. pericardial cavity
      iii. visceral layer (epicardium)
   b. myocardium
   c. endocardium
   d. apex of heart
   e. base of heart (at the valves)
   f. anterior interventricular sulcus
   g. posterior interventricular sulcus
   h. coronary sulcus
   i. "fibrous skeleton" of the heart

**j.** cardiac chambers

    **i.** right atrium and auricular appendage (auricle)

        1. interatrial septum

            a. fossa ovalis

            b. opening of the coronary sinus

        2. superior vena cava

        3. inferior vena cava

        4. areas of the sinoatrial (SA) and atrioventricular (AV) nodes

    **ii.** right ventricle

        1. tricuspid valve (left atrioventricular [AV] valve)

        2. chordae tendineae

        3. trabeculae carneae

        4. papillary muscles

        5. interventricular septum

        6. pulmonary (pulmonic) valve

        7. pulmonary trunk

            a. right pulmonary artery

            b. left pulmonary artery

    **iii.** left atrium

        1. auricular appendage (auricle)

        2. pulmonary veins

    **iv.** left ventricle

        1. bicuspid (mitral) valve (right AV valve)

        2. aortic valve

        3. ascending aorta

        4. aortic arch

            a. ligamentum arteriosum

**3.** coronary arteries

    **a.** right coronary artery (leads from the base of the ascending aorta)

        **i.** marginal artery

        **ii.** posterior descending artery [PDA] (posterior interventricular artery)

    **b.** left coronary artery (left main coronary artery)

        **i.** left anterior descending artery [LAD] (anterior interventricular artery)

        **ii.** circumflex artery

4. coronary veins

    **a.** great cardiac vein (paired with the LAD)

    **b.** middle cardiac vein (paired with the PDA)

    **c.** small cardiac vein

    **d.** coronary sinus

*Remember to show this pre-lab to the SSL instructor* **before leaving** *the lab. It will be marked off in the SSL book for grade credit towards your lab test.*

# 3

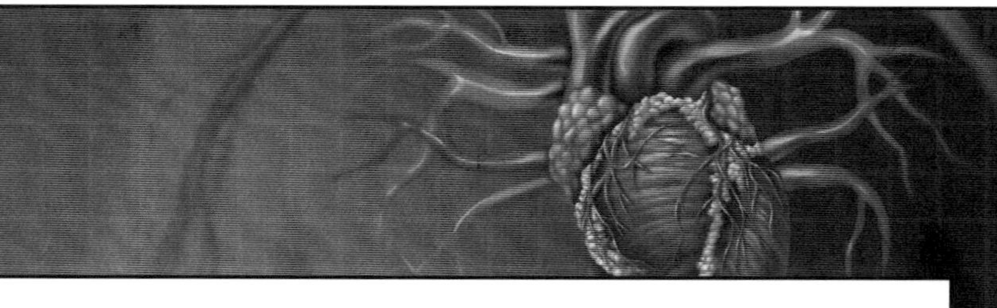

# THE HEART
## IN-LAB ACTIVITIES

Name: _______________________  Section: __________  Date: _________

## LEARNING OBJECTIVES

1. Identify major heart structures and know their functions.

2. Outline the flow of blood through the heart.

3. Become familiar with fetal circulation and heart maturation.

4. Explain and identify vessels involved in coronary circulation.

5. Describe microscopic structures of the heart muscles.

You will spend **2 hours or so** in lab at Forsyth Tech to complete the following activities. This amount of time allows you to complete the activities by using the torso model, using other models, and working with a lab partner.

# ACTIVITY 1: IDENTIFY MAJOR HEART STRUCTURES AND KNOW THEIR FUNCTIONS

*After reading through the Pre-Lab and online text for the heart, you will need enlarged heart models (if available), large pig heart to dissect, or other to find the anatomical structures.*

1. Study the model of the heart and label all structures shown on the figure below given on the anatomy list. (The model of the heart will be on our Lab Test.)

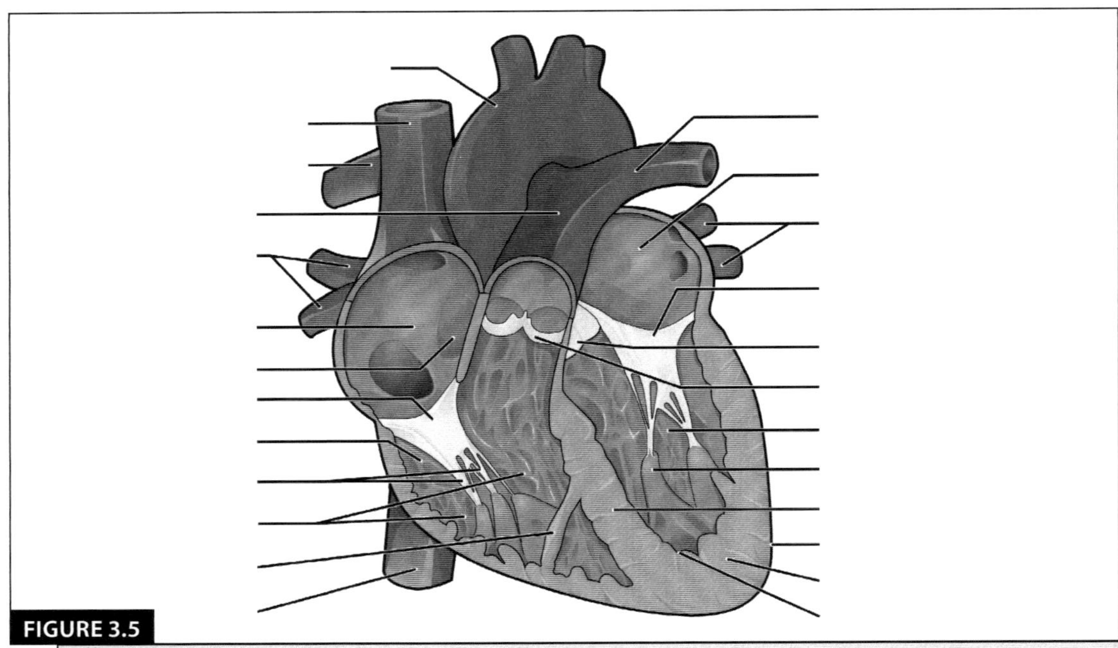

**FIGURE 3.5**

**The heart opened in a frontal section, showing the four chambers and the valves to label along with most other structures of the internal heart, as well as the layers of the heart wall.** Heart Anatomy [https://cnx.org/contents/FPtK1zmh@8.108:Y5T_wVSC@4/Heart-Anatomy] by OpenStax, (CC BY 4.0)

2. What are the three layers of the heart wall?

_________________________________________________________________

_________________________________________________________________

_________________________________________________________________

**You will continue to work on Activity 1 later on in lab with a pig heart dissection.**

# ACTIVITY 2: OUTLINE THE FLOW OF BLOOD THROUGH THE HEART

*You will need enlarged heart models (if available).*

1. Use the model diagram below and fill in the flow of blood through the heart, starting with the arrow at the right atrium. Do the same with your enlarged heart model.

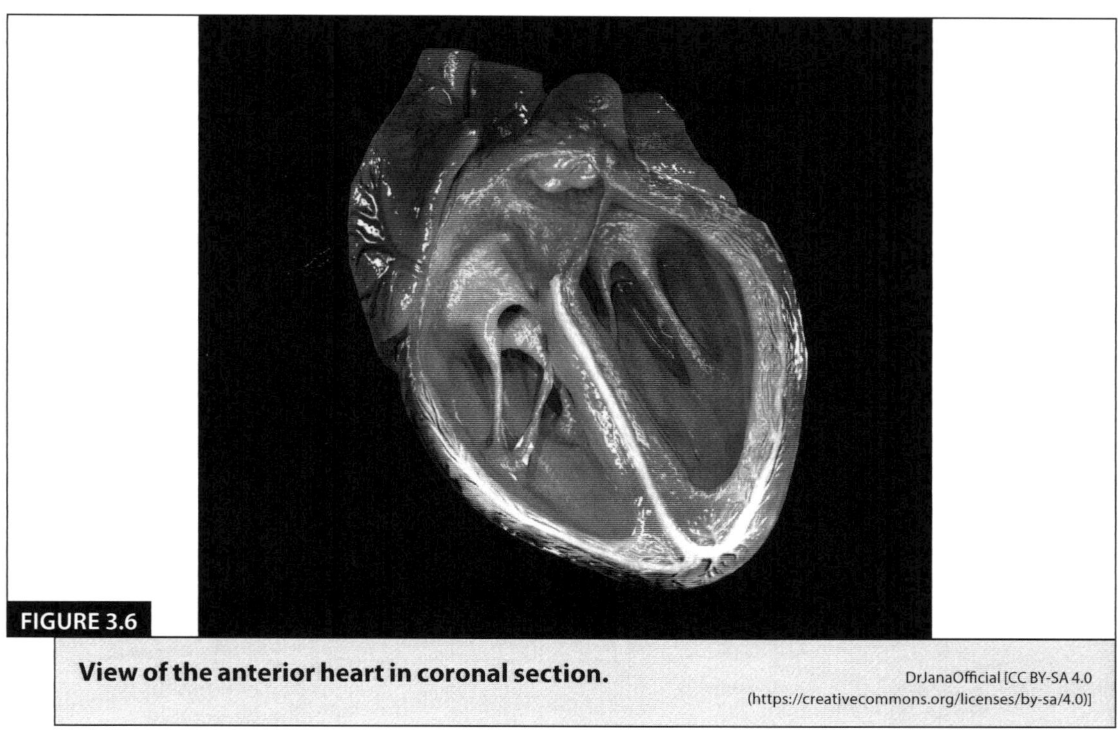

**FIGURE 3.6**

**View of the anterior heart in coronal section.**

DrJanaOfficial [CC BY-SA 4.0 (https://creativecommons.org/licenses/by-sa/4.0)]

- Where is the heart located in the thoracic cavity?

## ACTIVITY 3: BECOME FAMILIAR WITH FETAL CIRCULATION AND HEART MATURATION

*You will need enlarged heart models (if available).*

Locate the **fossa ovalis** on the heart model. It is an oval depression in the right atrium of the heart in the septum between right and left atrium. The fossa ovalis is the remnant of a thin fibrous sheet that covered the **foramen ovale** (oval opening) during fetal development.

1.  What is the advantage of having this opening while in utero? (**Hint:** oxygen supply)

2.  Also, find the remnant of the **ductus arteriosis** found only in the fetal heart.

    - Where is this located?

    - What is the remnant of this fetal structure called?

3.  In the heart outline below, draw and label the right atrium and ventricle along with the tricuspid valve and fossa ovalis. Find the valves in your dissected heart in the lab, too.

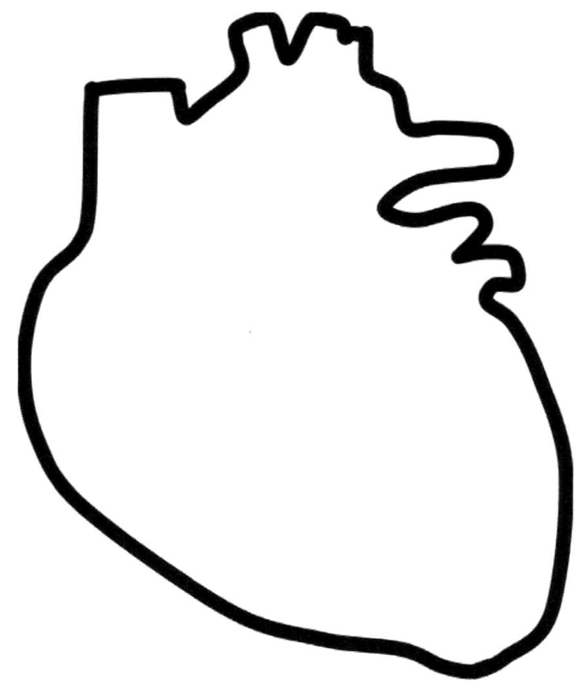

## ACTIVITY 4: EXPLAIN AND IDENTIFY VESSELS INVOLVED IN CORONARY CIRCULATION

*You will need enlarged heart models (if available).*

1. In the heart outlines below, draw and label the vessels of coronary circulation listed on the anatomy list. Color-code your drawing–red for artery and blue for vein.

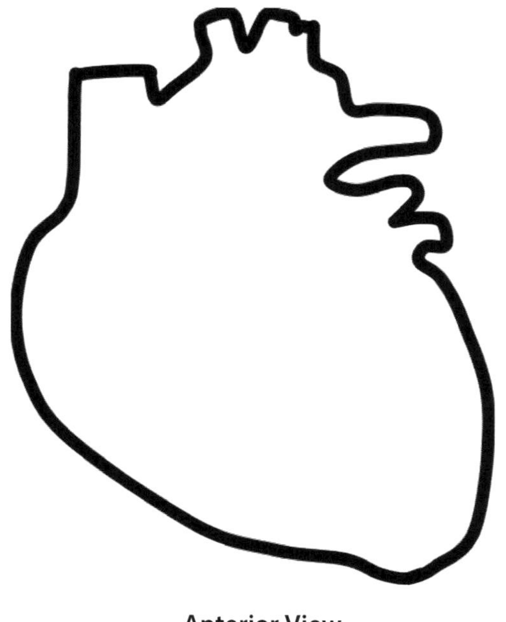

**Anterior View**

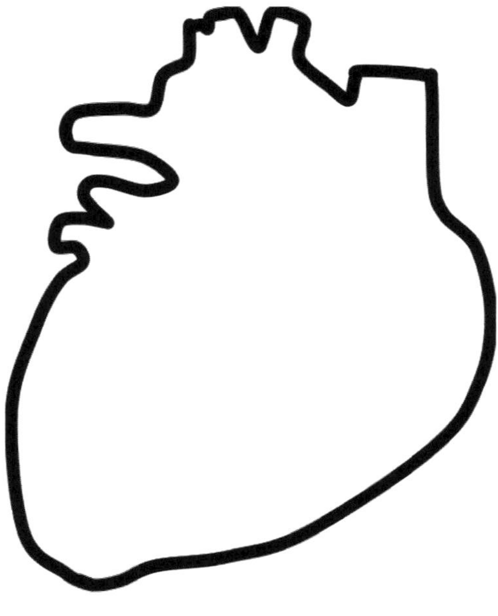

**Posterior View**

## ACTIVITY 5: DESCRIBE MICROSCOPIC STRUCTURES OF THE HEART MUSCLES

*You will need cardiac muscle histology slides and a compound light microscope (if available in lab or in a lab box).*

1.  View the cardiac muscle slide with **intercalated discs. Sketch** the muscle and discs below under high power.

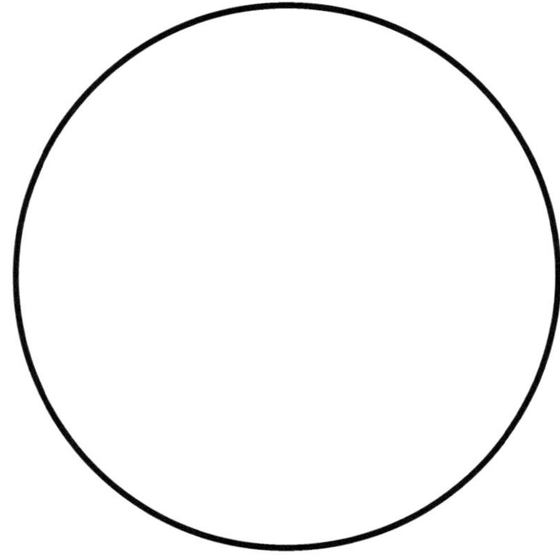

2.  Then list what two junctions the intercalated discs are formed from.

## ACTIVITY 1 CONTINUED: DISSECTION OF THE HEART

*Working with one or two partners, follow the directions below or follow the directions found in the lab room for the heart.*

### PART 1: EXTERNAL ANATOMY

*Some adapted from https://www.biologycorner.com/worksheets/heart_dissection.html*

1.  **Identify** the **anterior/ventral** and **posterior/dorsal** sides of the sheep's heart. Look closely and on the **anterior** side you will see an oblique line of blood vessels that divide the heart vertically; this line is called the interventricular sulcus. This dividing line of blood vessels is also seen on the posterior side but with different blood vessels. Identify the **right** and **left** sides of the heart. The half that has the **apex**

(pointed end) of the heart is the left side. If you are looking at the anterior side of the heart, then the apex should be on the opposite side from your left. Keep this in mind as you dissect and also look at the heart figures.

2. Locate the coronary arteries and veins that are on the surface of the heart. These may be covered by some fatty areas, but you can discern the blood vessels.

3. Find the flaps of tissue on the top of the heart. These ear-like flaps are called **auricles.** These sit on the two atrial chambers of the heart.

4. The anterior-most vessel is the pulmonary trunk. Place a straw in this vessel to mark its place.

5. Just posterior, or behind, the pulmonary trunk is the **aorta.** There may be a branch of the aorta called the **brachiocephalic artery,** but it is not always visible/attached.

6. Turn the heart so that you are looking at its dorsal or posterior side. Find the large opening at the top of the heart next to the right auricle. This is the **superior vena cava.** Place a probe in this vessel, and you can feel the inside of the **right atrium** when you reach in with your finger.

7. Locate another opening on the backside of the heart on the left side. This is the **pulmonary vein.** You can feel the inside of the left atrium by probing this opening with your finger.

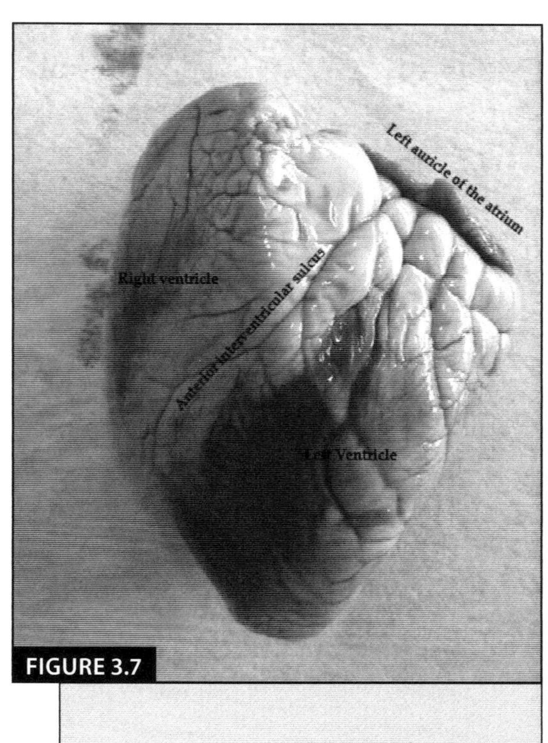


**FIGURE 3.7**

## Check Yourself

Make sure you know the location of each of the following before continuing to the internal anatomy of the heart:

- superior vena cava
- inferior vena cava
- aorta
- pulmonary trunk (and arteries)
- right atrium
- right ventricle
- left atrium
- pulmonary veins (seen in posterior view)
- left ventricle
- auricles
- coronary arteries & cardiac veins
- interventricular sulcus

8. Visually use the **lab models** to list which vessels connect to which **chambers.**

   - Pulmonary artery from the _______________________________________________

   - Pulmonary vein to the _________________________________________________

   - Aorta from the ________________________________________________________

   - Superior vena cava to the ______________________________________________

   - Left anterior descending artery from the ____________ (in the anterior sulcus)

   - Posterior descending artery from the ____________ (in the posterior sulcus)

## PART 2: INTERNAL ANATOMY

1. Use a scalpel to make an incision in the heart from the superior vena cava through the right atrium and ventricle along the **right margin** of the heart so that you can open the right side, viewing the **right atrium,** the **right ventricle,** and the **tricuspid valve** between.

2. Do you see the **chordae tendineae** inside the right ventricle? These look like chords ["heart strings"] that are attached to the tricuspid valve and the **papillary muscles.**

3. Cut open the **left atrium** and locate the **bicuspid valve** between the left atrium and ventricle. Cut along the margin and across the **left ventricle** anteriorly just below the top of the left ventricle and down the **interventricular septum.** Remove the lower-front portion of the wall. If the bicuspid valve is still in place, observe it before cutting through it next. Spread open the left ventricle, viewing the structures listed above.

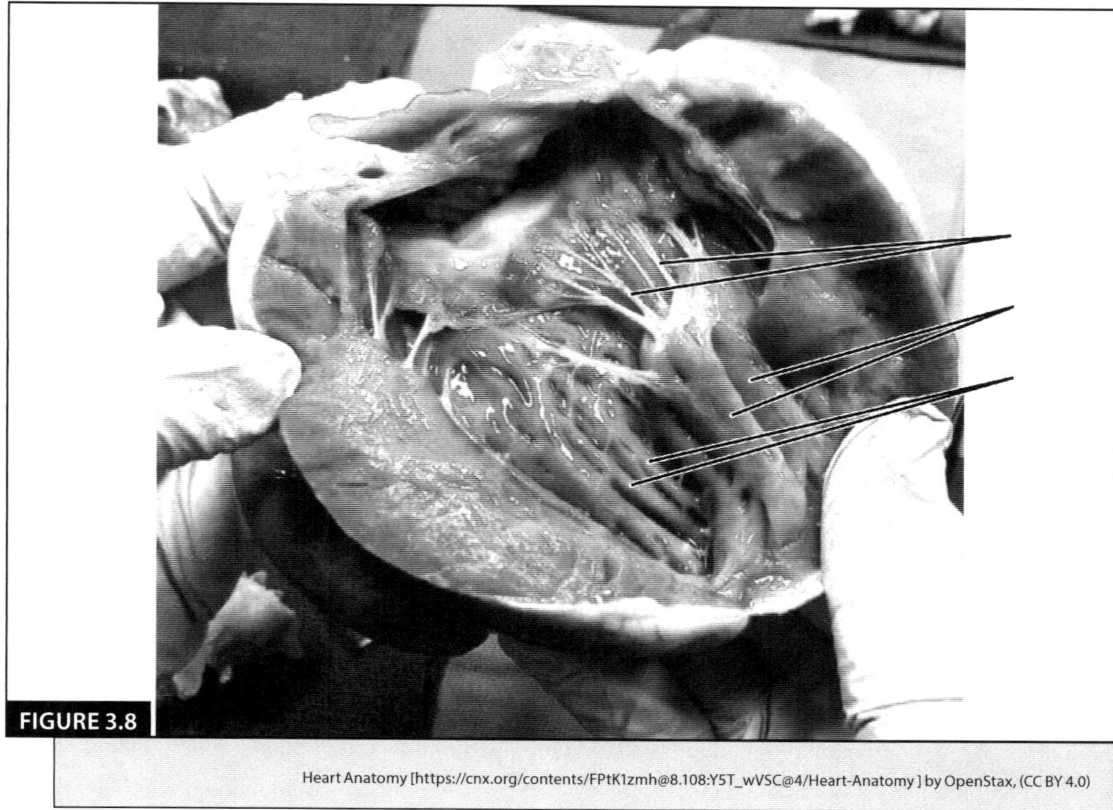



Heart Anatomy [https://cnx.org/contents/FPtK1zmh@8.108:Y5T_wVSC@4/Heart-Anatomy ] by OpenStax, (CC BY 4.0)

4. Remember the path of blood through the heart and **trace** it through the dissected heart from the right atrium.

- Compare the sheep or pig heart to the human heart by viewing the different heart models in the room.

- Be sure you can label the parts of the heart on the heart models in the lab room.

# ACTIVITY 6: REVIEW STRUCTURES OF THE HEART

*After this lab, you should have been able to find the following anatomical structures on the heart models available. Go through each structure twice, naming it out loud as you find it on the model(s).*

## ANATOMY LIST

1.  Identify the following **microscopic structures** on a slide of cardiac muscle tissue:

    **a.** cardiac muscle fibers (cardiocytes)

    **Note** the branching arrangement of the cardiocytes.

    **b.** intercalated discs

2.  Identify the following **cardiac structures** on models and diagrams:

    **a.** pericardium

     **i.**  parietal layer (fibrous and serous layers)

     **ii.**  pericardial cavity

     **iii.**  visceral layer (epicardium)

    **b.** myocardium

    **c.** endocardium

    **d.** apex of heart

    **e.** base of heart (at the valves)

    **f.**  anterior interventricular sulcus

    **g.** posterior interventricular sulcus

    **h.** coronary sulcus

    **i.**  "fibrous skeleton" of the heart

    **j.**  cardiac chambers

     **i.**  right atrium and auricular appendage (auricle)

      1.  interatrial septum

       a.  fossa ovalis

       b.  opening of the coronary sinus

      2.  superior vena cava

      3.  inferior vena cava

      4.  areas of the sinoatrial (SA) and atrioventricular (AV) nodes

      **ii.** right ventricle

         1. tricuspid valve (left atrioventricular [AV] valve)

         2. chordae tendineae

         3. trabeculae carneae

         4. papillary muscles

         5. interventricular septum

         6. pulmonary (pulmonic) valve

         7. pulmonary trunk

            a. right pulmonary artery

            b. left pulmonary artery

    **iii.** left atrium

         1. auricular appendage (auricle)

         2. pulmonary veins

    **iv.** left ventricle

         1. bicuspid (mitral) valve (right AV valve)

         2. aortic valve

         3. ascending aorta

         4. aortic arch

            a. ligamentum arteriosum

**3.** coronary arteries

  **a.** right coronary artery (leads from the base of the ascending aorta)

    **i.** marginal artery

    **ii.** posterior descending artery [PDA] (posterior interventricular artery)

  **b.** left coronary artery (left main coronary artery)

    **i.** left anterior descending artery [LAD] (anterior interventricular artery)

    **ii.** circumflex artery

**4.** coronary veins

  **a.** great cardiac vein (paired with the LAD)

  **b.** middle cardiac vein (paired with the PDA)

  **c.** small cardiac vein

  **d.** coronary sinus

*Be sure to get your completed work checked off by a member of the lab staff and then keep this hand out for your review.*

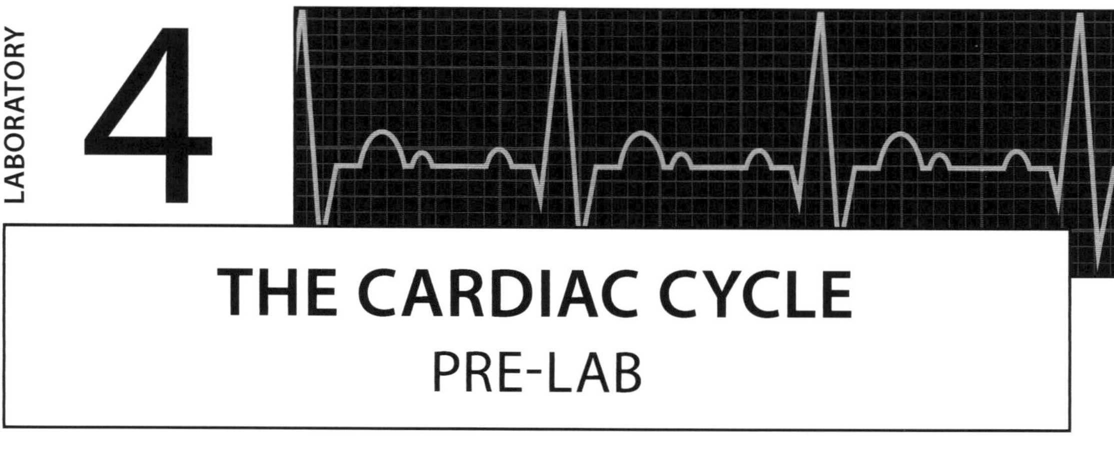

# THE CARDIAC CYCLE
## PRE-LAB

Name: _________________________     Section: ___________     Date: _________

## LEARNING OBJECTIVES

1. Detail the electrical conduction system of the heart.

2. Detail the cardiac events that cause the heart sounds and assess them by auscultation.

3. Investigate changes in the length and rate of one cardiac cycle in response to exercise.

4. Analyze the physical manifestation of electrical energy in an electrocardiogram (ECG).

*Checklist to complete* **before entering** *the science skills lab (SSL):*

☐ Actively read this packet of information.

☐ Complete the charts, tables or labeling and answer questions using your own words.

☐ Complete the electronic digital pre-lab quiz on Bb by Sunday.

☐ Review the attached anatomy list and take it to lab with you. Jot down key descriptive identifying words that aid in your lab test preparations.

## ACTIVITY 1: DETAIL THE ELECTRICAL CONDUCTION SYSTEM OF THE HEART

*Read the content in your text and the introductory paragraphs below. Then, complete Activity 1 in lab.*

The heart is capable of generating its own electrical impulse, triggered by the pacemaker cells, as part of the cardiac conduction system. The components of the cardiac conduction system include the sinoatrial node, intermodal pathways, the atrioventricular node, the atrioventricular bundle, the atrioventricular bundle branches, and the purkinje cells.

Label these components on Figure 4.1 below.

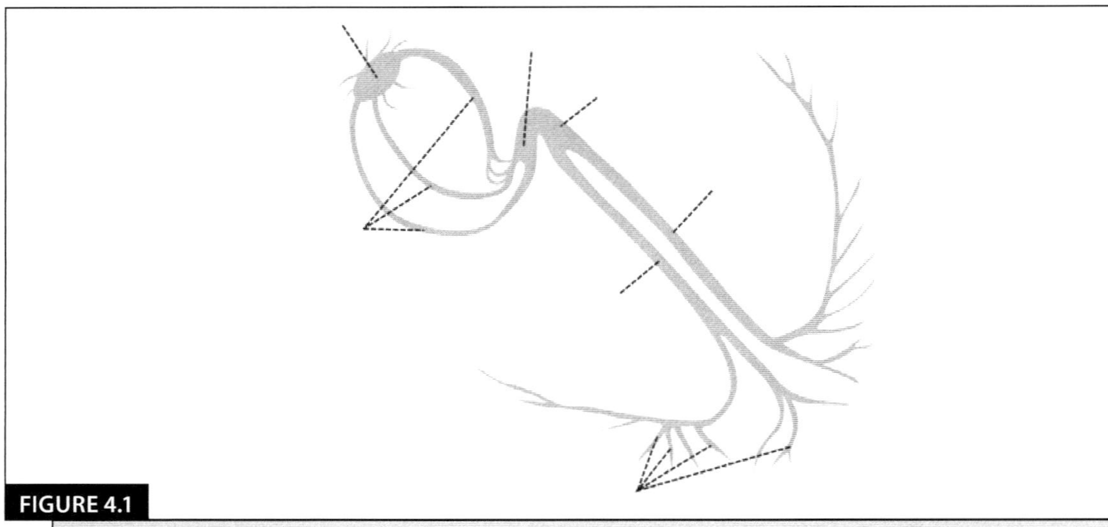

**FIGURE 4.1**

**This figure shows the physical image of the conduction system of the heart and its location through the heart.** This system distributes the electrical energy that goes through the heart. An ECG is the functional image of the electrical activity of the heart. Madhero88
[CC BY-SA 3.0 (https://creativecommons.org/licenses/by-sa/3.0)]

Normal cardiac rhythm is established by the **sinoatrial (SA) node,** a specialized clump of myocardial conducting cells located in the superior and posterior walls of the right atrium in close proximity to the orifice of the superior vena cava. The SA node has the highest inherent rate of depolarization and is known as the **pacemaker** of the heart. It initiates the **sinus rhythm** or the normal electrical pattern followed by contraction of the heart. The electrical event, the wave of depolarization, is the trigger for muscular contraction.

This impulse spreads from its initiation in the SA node throughout the atria through specialized **internodal pathways,** to the atrial myocardial contractile cells and the atrioventricular node. The **atrioventricular (AV) node** is a second clump of specialized myocardial conductive cells, located in the inferior portion of the right atrium within the atrioventricular septum. The septum prevents the impulse from spreading directly to the ventricles without passing through the AV node. There is a critical pause that allows the atrial cardiomyocytes to complete their contraction before the AV node depolarizes and transmits the impulse to the atrioventricular bundle. The **atrioventricular bundle,** or **bundle of His,** proceeds through the interventricular septum before dividing into two **atrioventricular bundle branches,** commonly called the left and right bundle branches. The bundle branches terminate into **Purkinje fibers** that spread the impulse to the myocardial contractile cells in the ventricles. They extend throughout the myocardium from the apex of the heart toward the atrioventricular septum and upward toward the base of the heart.

The total time elapsed from the initiation of the impulse in the SA node until depolarization of the ventricles is approximately 225 ms. For the figure below, place the number of the image by the statement that describes what event is occurring. Use the figure to label the number of the events at given each stage.

1. _______ Impulse travels through the AV bundle and bundle branches to the Purkinje fibers

2. _______ The SA node and the remainder of the conduction system are at rest.

3. _______ Impulse spreads to the contractile fibers of the ventricle

4. _______ Delay at the AV node

5. _______ Ventricular contraction begins.

6. _______ The SA node initiates the action potential.

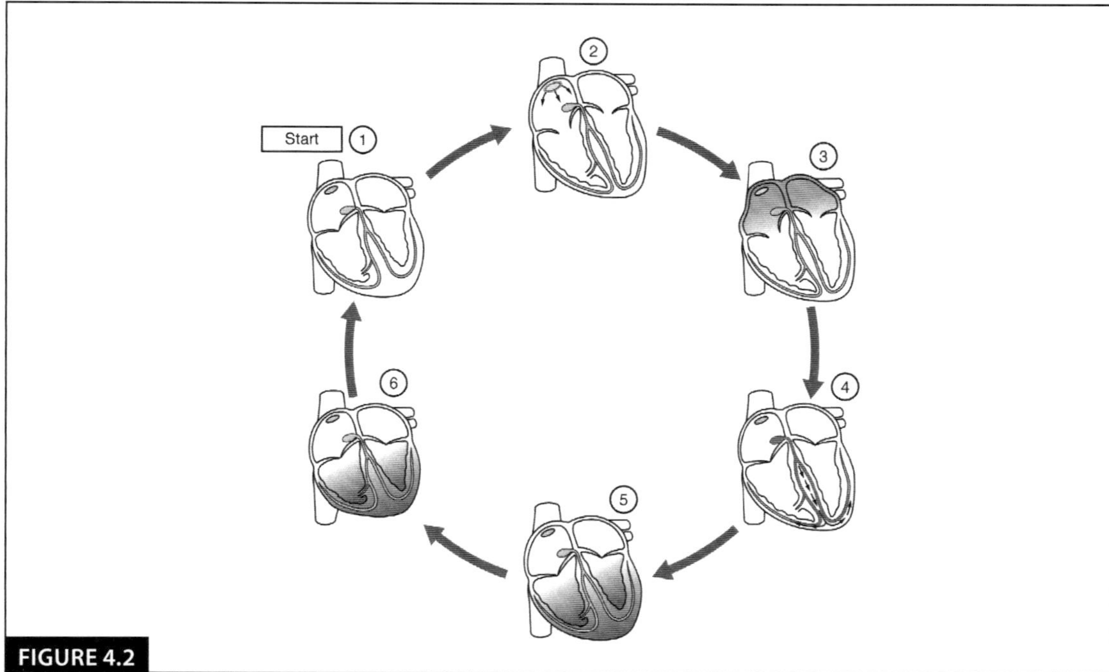


**FIGURE 4.2**

**The total cycle of the initiation of the impulse in the SA node until depolarization of the ventricles**

OpenStax College [CC BY 3.0 (https://creativecommons.org/licenses/by/3.0)]

## ACTIVITY 2: PERFORM AUSCULTATION OF HEART SOUNDS AND DETAIL THE CARDIAC EVENTS THAT CAUSE THESE SOUNDS

*Read the content in your text and the introductory paragraphs below. Then, complete Activity 2 in lab.*

One of the simplest, yet effective, diagnostic techniques applied to assess the state of a patient's heart is auscultation using a stethoscope. In a normal, healthy heart, there are only two audible **heart sounds: S1** and **S2.** S1 is the sound created by the closing of the atrioventricular valves (tricuspid, bicuspid, or mitral) during ventricular contraction and is normally described as a "lub," or first heart sound. The second heart sound, S2, is the sound of the closing of the semilunar valves (aortic and pulmonary) during ventricular diastole and is described as a "dub."

The term **murmur** is used to describe an unusual sound coming from the heart that is caused by the turbulent flow of blood as it passes through insufficient heart valves, or blood flow returning to the preceding chamber, which is abnormal. This is called blood regurgitation. During auscultation, it is common practice for the clinician to ask the patient to breathe deeply. This procedure not only allows for listening to airflow, but it may also amplify heart murmurs. Inhalation increases blood flow into the right side of the heart and may increase the amplitude of right-sided heart murmurs. Expiration partially restricts blood flow into the left side of the heart and may amplify left-sided heart murmurs.

- What causes the "lub" sound?

________________________________________________________________

________________________________________________________________

- What causes the "dub" sound?

________________________________________________________________

________________________________________________________________

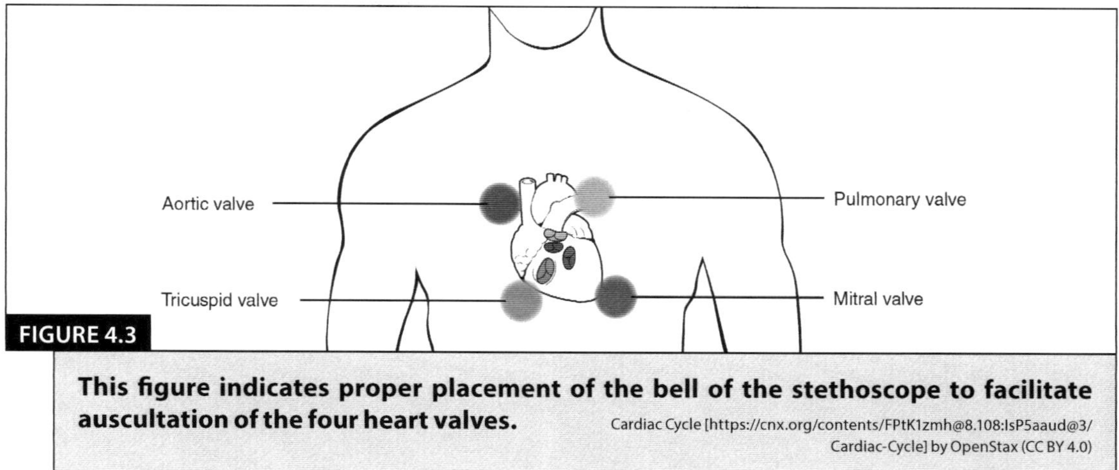


**FIGURE 4.3**

**This figure indicates proper placement of the bell of the stethoscope to facilitate auscultation of the four heart valves.** Cardiac Cycle [https://cnx.org/contents/FPtK1zmh@8.108:IsP5aaud@3/Cardiac-Cycle] by OpenStax (CC BY 4.0)

Sounds will be best heard on the side of the direction of blood flow as it passes through the heart valves. This means distal to the valve in the direction of blood flow.

## ACTIVITY 3: INVESTIGATE CHANGES IN THE LENGTH OF ONE CARDIAC CYCLE IN RESPONSE TO EXERCISE

*Read the content in your text and the introductory paragraphs below. Then, complete Activity 3 in lab.*

The period of time that begins with contraction of the atria and ends with ventricular relaxation is known as the **cardiac cycle.**

The period of contraction that the heart undergoes while the chambers pump blood into circulation is called **systole or ejection.** These terms are synonyms: systole = contraction = ejection = emptying.

The period of relaxation that occurs as the chambers fill with blood is called **diastole** or filling. These terms are synonyms: diastole = relaxation = injection = filling.

Both the atria and ventricles undergo systole and diastole at sequentially different times, and it is essential that these components be carefully regulated and coordinated to ensure blood is pumped efficiently to the body. The length of one cardiac cycle in one minute is calculated by dividing 60 by the number of heart beats. This means that if your heart rate changes due to activity, caffeine, meditation, etc., the length of your cardiac cycle will change as well.

Before going to lab, name the five phases of the cardiac cycle and give a brief description of the event that occurs during each phase in your own words.

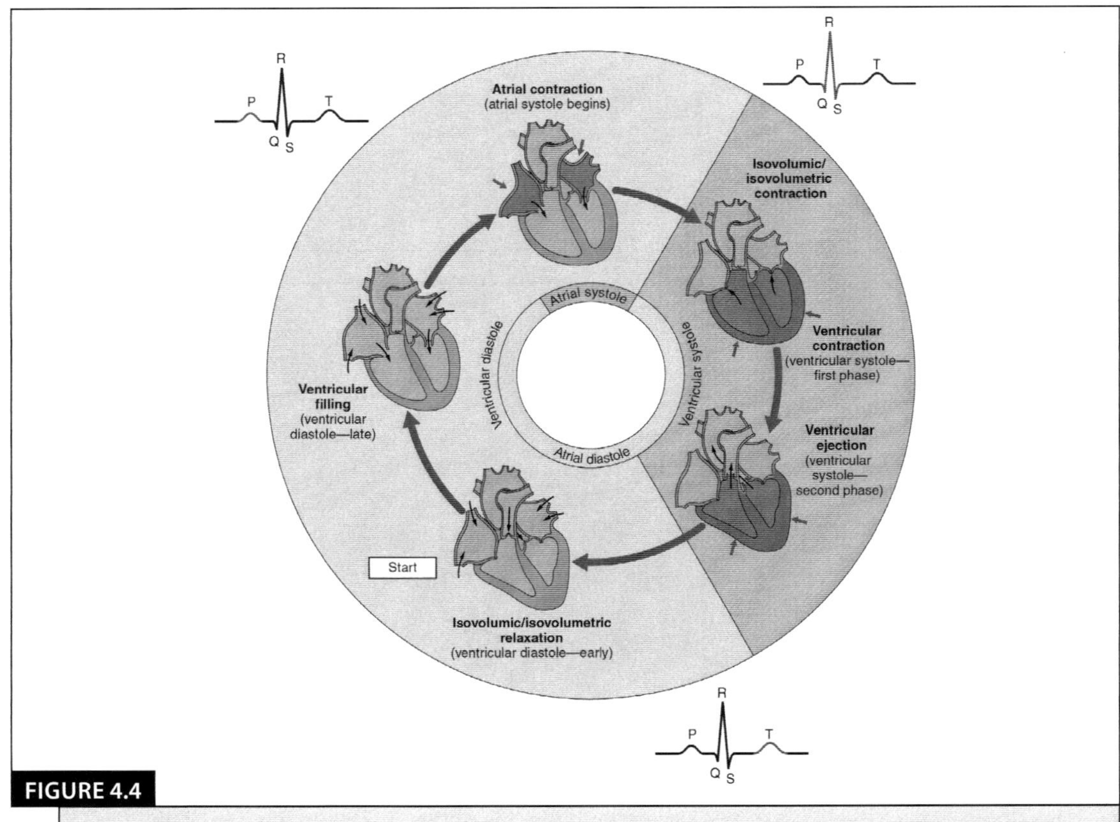


**FIGURE 4.4**

**The pie chart indicates the five phases of the cardiac cycle during atrial systole and diastole and during ventricular systole and diastole.** OpenStax College [CC BY 3.0 (https://creativecommons.org/licenses/by/3.0)]

1. _______________________________________________

2. _______________________________________________

3. _______________________________________________

4. _______________________________________________

5. _______________________________________________

# ACTIVITY 4: ANALYZE THE PHYSICAL MANIFESTATION OF ELECTRICAL ENERGY IN AN ECG

## ELECTROCARDIOGRAM (ECG OR EKG)

By careful placement of surface electrodes on the body, it is possible to record the complex, compound electrical signal of the heart. This tracing of the electrical signal is the **electrocardiogram (ECG)**. Careful analysis of the ECG reveals a detailed picture of both normal and abnormal heart function and is an indispensable clinical diagnostic tool. The standard electrocardiograph (the instrument that generates an ECG) uses three, five, or twelve leads. The greater the number of leads an electrocardiograph uses, the more information the ECG provides. The twelve-lead electrocardiograph uses ten electrodes placed in standard locations on the patient's skin–see below.

A normal ECG tracing is presented in Figure 4.6 on the next page. Each component, segment, and interval is labeled and corresponds to important electrical events, demonstrating the relationship between these events and contraction in the heart. There are five prominent points on the ECG: the P wave, the QRS complex, and the T wave.

The small **P wave** represents the depolarization of the atria. The atria begins contracting approximately 25 ms after the start of the P wave. The large **QRS complex** represents the depolarization of the ventricles, which requires a much stronger electrical signal because of the larger size of the ventricular cardiac muscle. The ventricles begin to contract as the QRS reaches the peak of the R wave. Lastly, the **T wave** represents the repolarization of the ventricles. The repolarization of the atria occurs during the QRS complex, which masks it on an ECG (non-recorded on the wave trace).

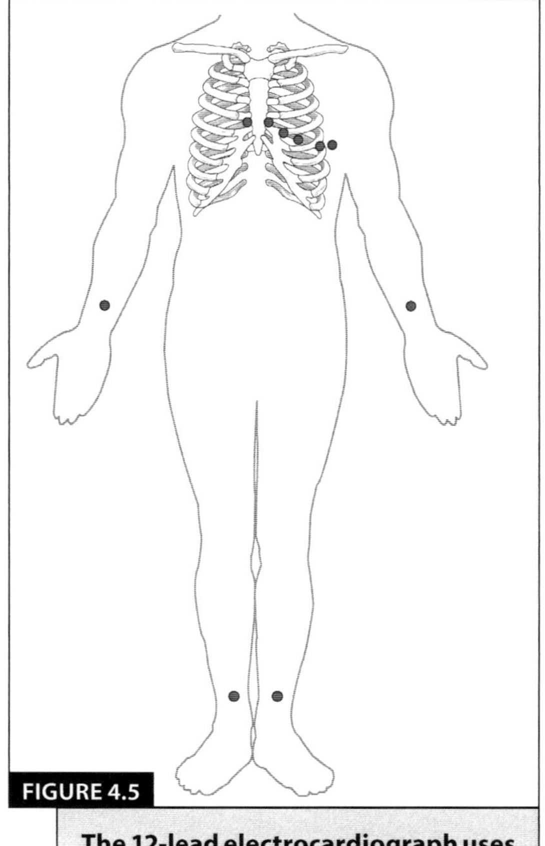

**FIGURE 4.5**

**The 12-lead electrocardiograph uses ten electrodes placed in standard locations on the patient's skin.**
OpenStax College [CC BY 3.0 (https://creativecommons.org/licenses/by/3.0)]

**Segments** are defined as the regions between two waves. **Intervals** include one segment plus one or more waves. For example, the PR segment begins at the end of the P wave and ends at the beginning of the QRS complex. The PR interval starts at the beginning of the P wave and ends with the beginning of the QRS complex. The PR interval is more clinically relevant, as it measures the duration from the beginning of atrial depolarization (the P wave) to the initiation of the QRS complex. Should there be a delay in passage of the impulse from the SA node to the AV node, it would be visible in the PR interval.

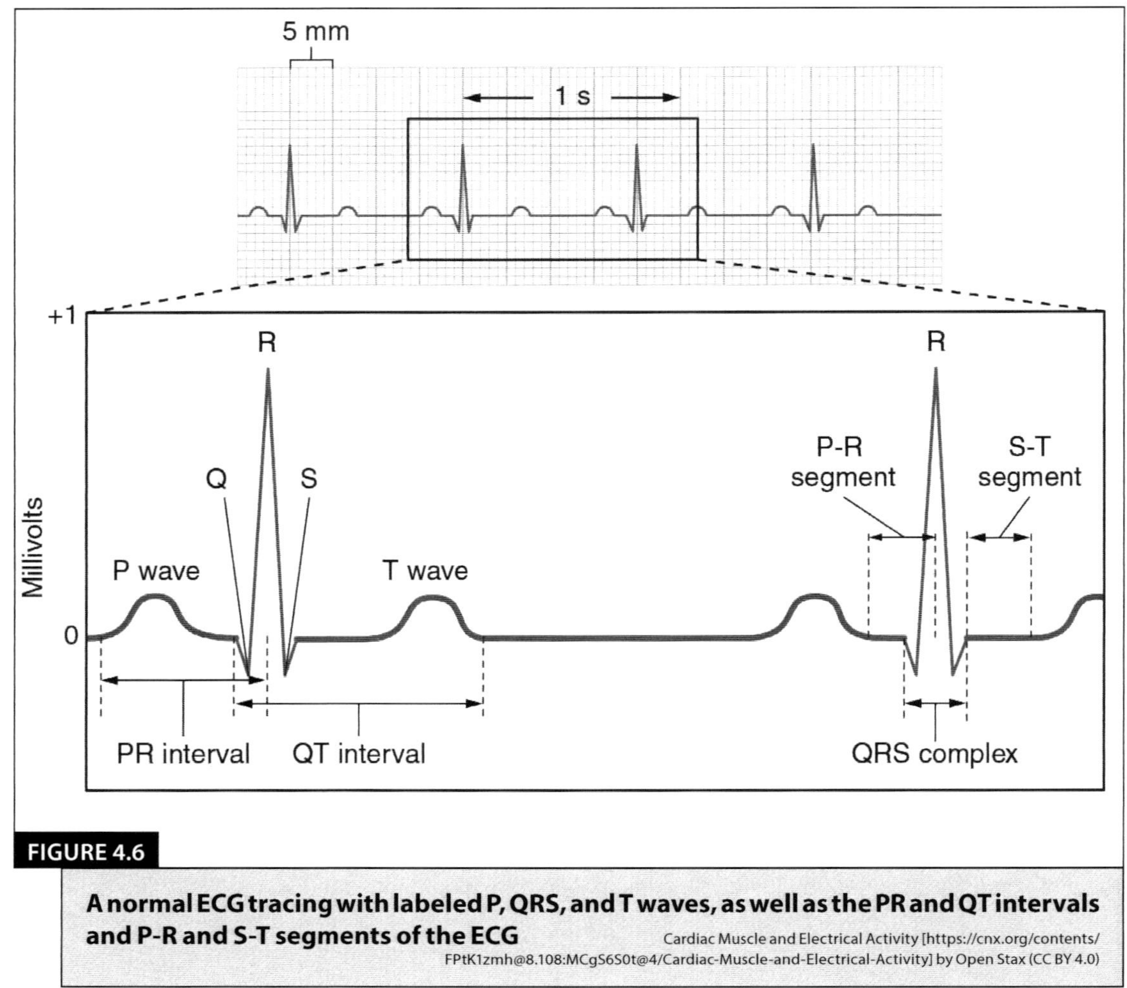


**FIGURE 4.6**

**A normal ECG tracing with labeled P, QRS, and T waves, as well as the PR and QT intervals and P-R and S-T segments of the ECG**    Cardiac Muscle and Electrical Activity [https://cnx.org/contents/FPtK1zmh@8.108:MCgS6S0t@4/Cardiac-Muscle-and-Electrical-Activity] by Open Stax (CC BY 4.0)

In lab, you can perform a simple ECG using a three-lead to view the electrical signal of the heart. Know the regions of the ECG above in order to label these on your or your lab partner's ECG.

*Checklist to complete* **before entering** *the science skills lab (SSL):*

☐ Actively read this packet of information.

☐ Complete the charts, tables or labeling and answer questions using your own words.

☐ Complete the electronic digital pre-lab quiz on Bb by Sunday.

☐ Review the attached anatomy list and take it to lab with you. Jot down key descriptive identifying words that aid in your lab test preparations.

## ANATOMY LIST

1. Identify the following components of the electrical conduction system of the heart on models and diagrams:
   a.  sinoatrial node
   b.  atrioventricular node
   c.  AV bundle (bundle of His)
   d.  bundle branches (right and left)
   e.  Purkinje fibers

2. Identify the parts and rhythms of an ECG on diagrams and strips:
   a.  wave
   b.  PQ interval
   c.  QRS complex
   d.  ST segment
   e.  T wave
   f.  QT interval
   g.  normal sinus rhythm

   In lab you may see the following on ECG strips:

   a.  bradycardia
   b.  tachycardia
   c.  ventricular fibrillation
   d.  atrial fibrillation
   e.  heart blocks (first- third degree)

*Remember to show this pre-lab to the SSL instructor* **before leaving** *the lab. It will be marked off in the SSL book for grade credit towards your lab test.*

# 4

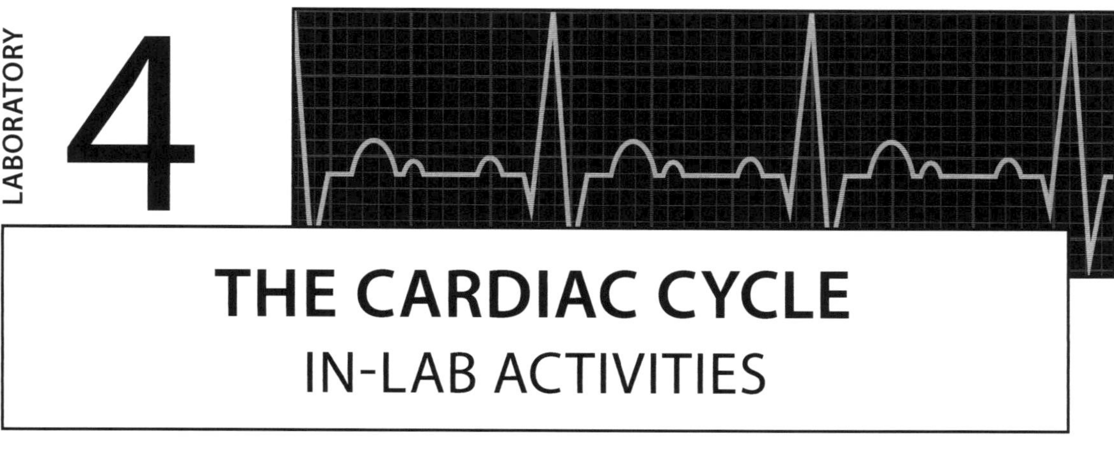

# THE CARDIAC CYCLE
## IN-LAB ACTIVITIES

Name: _________________________  Section: _________  Date: _________

## LEARNING OBJECTIVES

1. Detail the electrical conduction system of the heart.

2. Detail the cardiac events that cause the heart sounds and assess them by auscultation.

3. Investigate changes in the length and rate of one cardiac cycle in response to exercise.

4. Analyze the physical manifestation of electrical energy in an electrocardiogram (ECG).

You will spend **2–3 hours or so** in lab to complete the following activities. This amount of time allows you to complete the activities by using the torso model, using other models, and working with a lab partner.

## ACTIVITY 1: DETAIL THE ELECTRICAL CONDUCTION SYSTEM OF THE HEART

*Label the parts of the electrical conduction system of the heart in the figure below.*

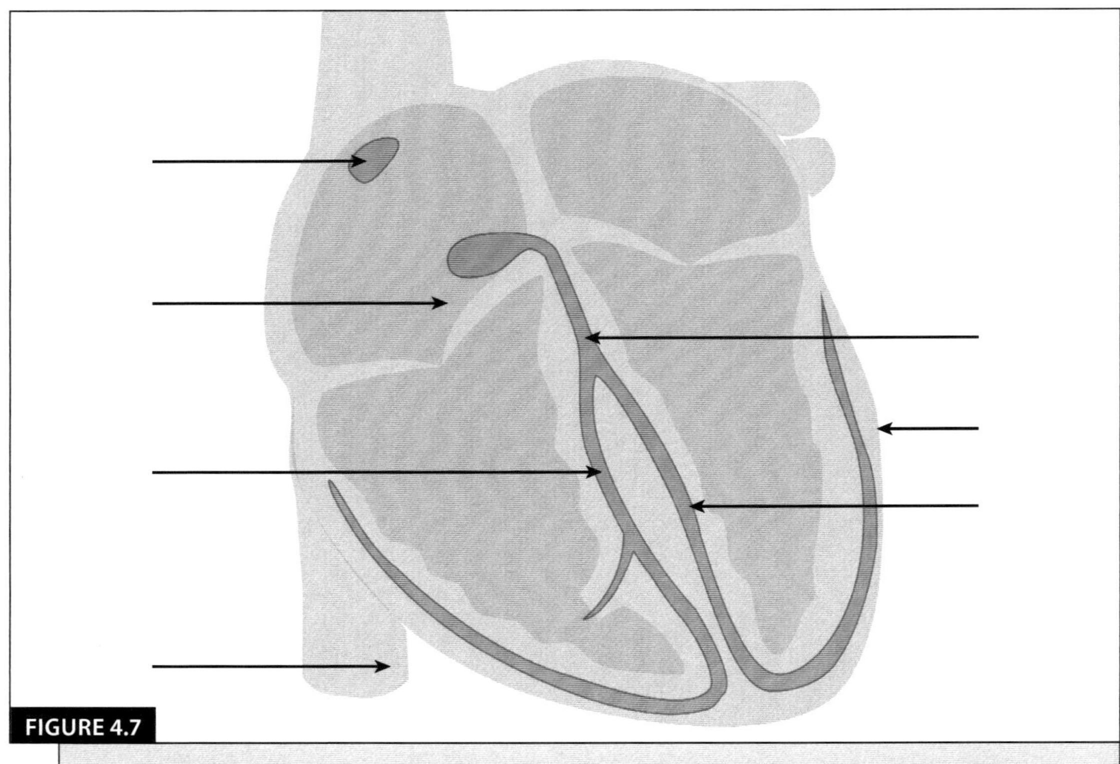

**FIGURE 4.7**

**The electrical conduction system of the heart to be labeled by students in the lab.**
U Bhalraam [CC BY-SA 4.0 (https://creativecommons.org/licenses/by-sa/4.0)]

# ACTIVITY 2: PERFORM AUSCULTATION OF HEART SOUNDS AND DETAIL THE CARDIAC EVENTS THAT CAUSE THESE SOUNDS.

*You will need a stethoscope and a partner for this activity (if available).*

1. Grab a partner and place the bell of the stethoscope on each of the four locations denoted below. Record what you hear in the space provided below.

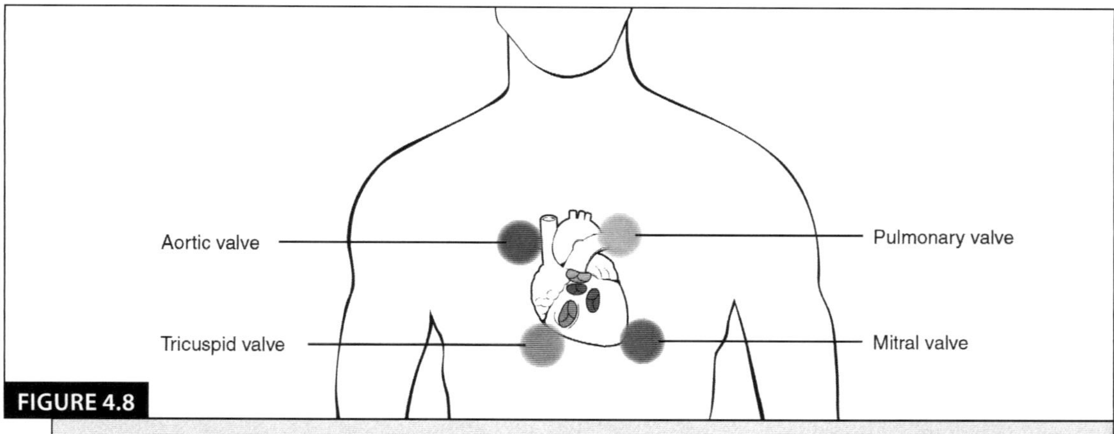


**FIGURE 4.8**

**Proper placement of the bell of the stethoscope facilitates auscultation.** At each of the four locations on the chest, a different valve can be heard. Cardiac Cycle [https://cnx.org/contents/ FPtK1zmh@8.108:IsP5aaud@3/Cardiac-Cycle] by OpenStax (CC BY 4.0)

- Aortic valve closure observations:

  _______________________________________________________________

- Tricuspid valve closure observations:

  _______________________________________________________________

- Pulmonary valve closure observations:

  _______________________________________________________________

- Mitral valve closure observations:

  _______________________________________________________________

2. At rest, was the "lub" or "dub" sound louder? Why do you think that is?

  _______________________________________________________________

  _______________________________________________________________

  _______________________________________________________________

## ACTIVITY 3: INVESTIGATE CHANGES IN THE LENGTH OF ONE CARDIAC CYCLE IN RESPONSE TO EXERCISE

*You will need a partner for this activity (if available).*

1.  Begin by locating the radial and carotid pulses on yourself and on your lab partner.

    a.  Radial pulse: Using your index and middle fingers, locate the groove between the radius and the tendon on the anterior surface of the wrist. Do not use your thumb–it has its own noticeable pulse and will skew the results.

    b.  Carotid pulse: Using the same digits, locate the groove adjacent to the larynx between the thyroid cartilage and sternocleidomastoid muscle.

2.  At rest in a sitting position, have your partner take your heart rate at the radial pulse while you take your own at the contralateral carotid pulse. Count the pulses in 15 seconds and then multiply that value by 4. Take the average of the two readings and record your value in the table below.

3.  Exercise by running in place for two minutes.

4.  Immediately following exercise return to the sitting position and take the pulses again (radially and at the carotid). Record your averaged value in the table below.

5.  Continue to take pulse measurements every minute until the heart rate returns to the resting value. Record how long the recovery took in the table below.

6.  Once the recovery is complete, calculate the length of the resting and post-exercise cardiac cycle by dividing 60 by the number of heart beats.

7.  For the resting length of the cardiac cycle, calculate the amount of time the cycle was in systole versus diastole. This can be done by simply multiplying the length of the cardiac cycle by 1/3 to find the amount of time in systole, and then multiplying the length of the cardiac cycle by 2/3 to find the amount of time in diastole. Record your values in the table below.

8.  Switch roles and repeat the experiment with your partner now exercising. Record the results in the table below.

| | RESTING HR (BPM) | POST EXERCISE HR (BPM) | # OF MINUTES FOR RECOVERY | RESTING CARDIAC CYCLE LENGTH (SEC) | RESTING SYSTOLE (SEC) | RESTING DIASTOLE (SEC) | POST-EXERCISE CARDIAC CYCLE LENGTH (SEC) |
|---|---|---|---|---|---|---|---|
| Subject 1 | | | | | | | |
| | | | | | | | |
| Subject 2 | | | | | | | |

- How does exercise affect the length of the cardiac cycle?

- Why is it inaccurate to calculate the length of systole and diastole after exercise with a formula like we were able to do at rest?

- What are some factors that affect recovery time?

## ACTIVITY 4: ANALYZE THE PHYSICAL MANIFESTATION OF ELECTRICAL ENERGY IN AN ECG

### ELECTROCARDIOGRAM (ECG OR EKG)

An electrocardiography is a method to detect the electrical activity of the heart. Large numbers of cardiac muscle cells change their membrane potential, the electrical potential (voltage) across their plasma membranes, in short periods of time. Though the voltages created by each cell are small, the activity of large numbers of cells acting all at once creates enough voltage to be detected on the surface of the skin. When metal wires that conduct electricity are attached to the skin in standardized arrangements, a standardized pattern of voltage change is observed that correlates with the voltage changes taking part in various regions of the heart.

A typical pattern of voltage changes observed using leads placed on the skin and recording voltage changes is shown in Figure 4.9. This is an electrocardiogram (ECG). The peaks and valleys are called waves and are named in alphabetical order, P, Q, R, S, and T waves. Each wave represents electrical activity in different regions of the heart, and the duration of the waves and the time between waves represents the conduction of electricity from one region of the heart to the next, as outlined in your laboratory manual.

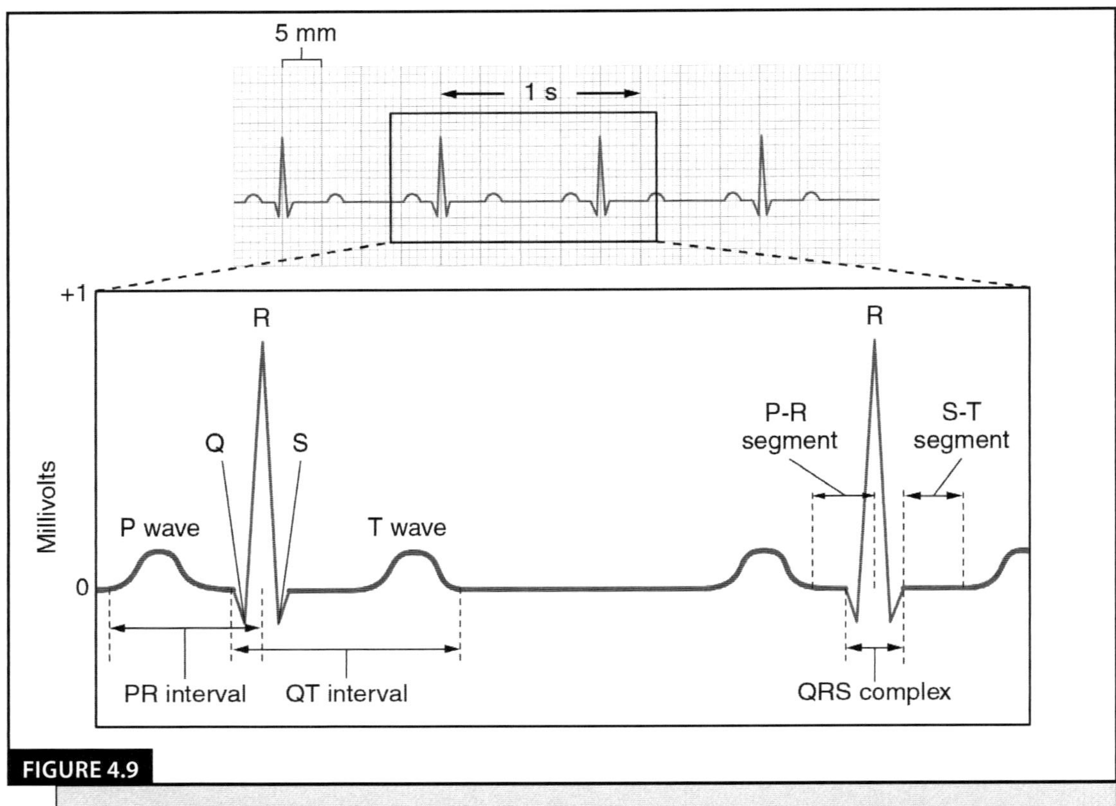


**FIGURE 4.9**

**Electrocardiogram.** A normal tracing shows the P wave, QRS complex, and T wave. Also indicated are the PR, QT, QRS, and ST intervals, plus the P-R and S-T segments. Cardiac Muscle and Electrical Activity [https://cnx.org/contents/FPtK1zmh@8.108:MCgS6S0t@4/Cardiac-Muscle-and-Electrical-Activity] by Open Stax (CC BY 4.0)

1. Detail the electrical activities that cause each of the waveforms and intervals from a standard ECG listed below.

   - P wave:_______________________________________________________
   - PR interval:___________________________________________________
   - QRS complex:__________________________________________________
   - ST segment:___________________________________________________
   - T wave:_______________________________________________________

2. Label the waveforms and intervals listed in question number two on the blank ECG below.

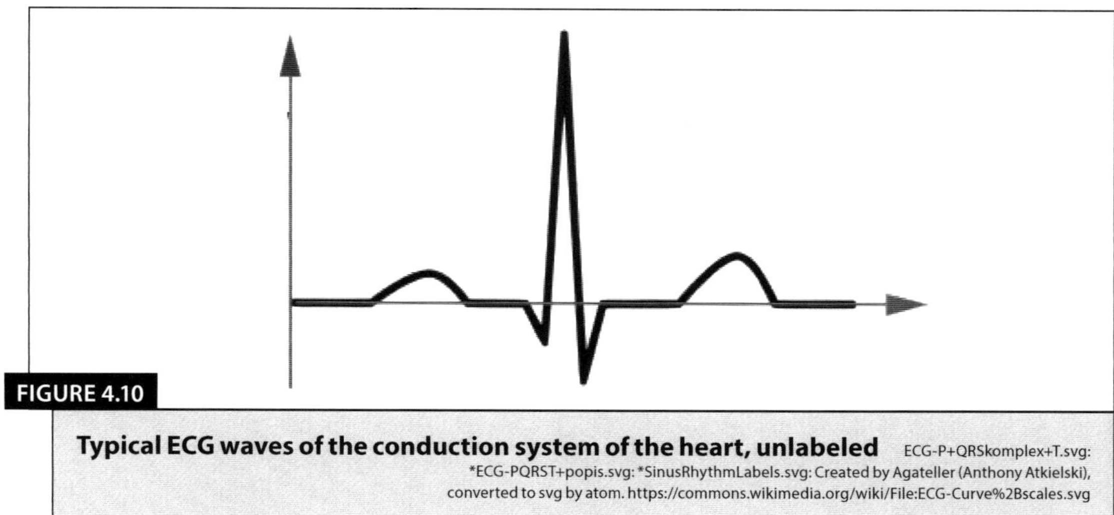



**Typical ECG waves of the conduction system of the heart, unlabeled** ECG-P+QRSkomplex+T.svg: *ECG-PQRST+popis.svg: *SinusRhythmLabels.svg: Created by Agateller (Anthony Atkielski), converted to svg by atom. https://commons.wikimedia.org/wiki/File:ECG-Curve%2Bscales.svg

3. Read the three ECG strips below and determine the heart rate (in beats per minute). Then, determine which strip is normal sinus rhythm. Bradycardia? Tachycardia?

   Directions for how to read an ECG strip are as follows:

   a. One large square (5 mm) = 0.2 seconds so 30 large squares = 6 seconds.

   b. Count the number of R waves in the 6 second section of the ECG strip.

   c. Multiply your number by 10 (There are 60 seconds in a minute and you want the unit beats per minute. 6 x 10 = 60, so your number x 10 = beats per minute.)

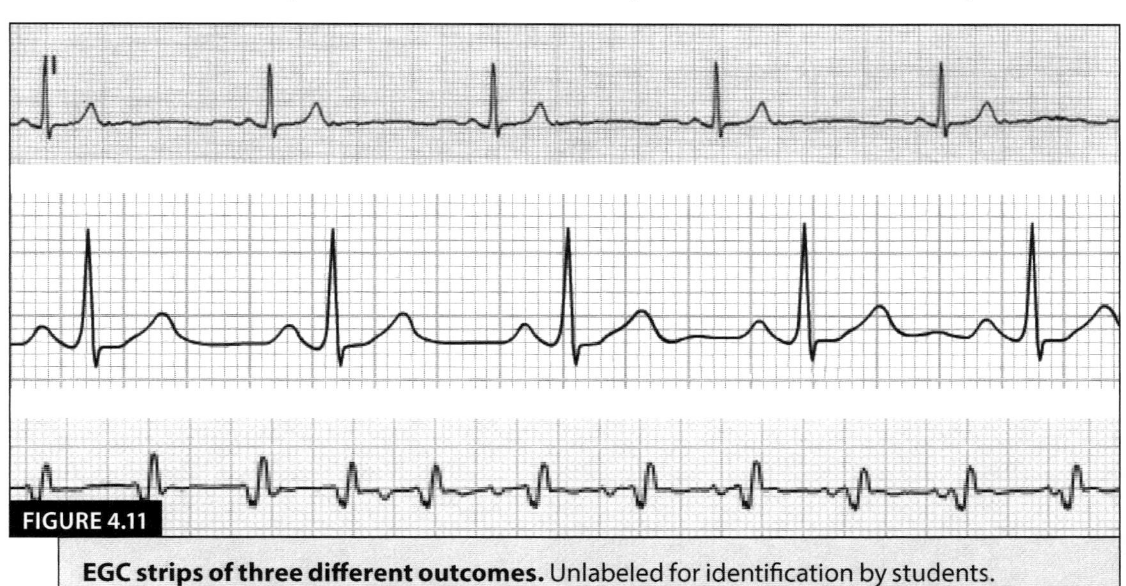



**EGC strips of three different outcomes.** Unlabeled for identification by students.
(Top) Glenlarson [Public domain] https://commons.wikimedia.org/wiki/File:Lead_II_rhythm_generated_sinus_bradycardia.JPG. (Middle) Andrewmeyerson [CC BY-SA 3.0 (https://creativecommons.org/licenses/by-sa/3.0)] https://commons.wikimedia.org/wiki/File:Normal_Sinus_Rhythm_Unlabeled.jpg  (Bottom) Jer5150 [CC BY-SA 3.0 (https://creativecommons.org/licenses/by-sa/3.0)] https://commons.wikimedia.org/wiki/File:Double_tachycardia_with_captures.png

## ACTIVITY 5: PERFORM THE ECG AVAILABLE IN THE LAB USING THE VERNIER HANDHELD QUESTS OR OTHER MECHANISM TO RECORD AN ECG BETWEEN YOU AND A PARTNER

*Materials and methods for Vernier Handheld Quests are below. [If you use the Vernier Mini Lab Quest, then use the directions found in lab.]*

A LabQuest (Vernier) handheld computer will be used to display and record the ECG. The separate ECG Sensor is plugged into any of the receptacles marked "CH1" through "CH4" located at the top of the LabQuest computer.

The adhesive electrical contacts which are placed on the skin are blue and about 2 cm × 2 cm with a small black non-adhesive tab on one edge. Clean your wrists with soap and water and dry thoroughly. Peel three contacts from their backing material and place, with some pressure, one on the anterior surface of each wrist and on the posterior surface of the right wrist.

There are three alligator clips coming from the ECG Sensor. Connect the green clip to the tab on the adhesive contact at the anterior surface of the right wrist, the black clip to the tab on the posterior surface of the right wrist, and the red clip to the tab on the adhesive contact on the left wrist.

Turn on the LabQuest by pressing the button at the upper left corner of the device. Keeping the ECG sensor away from all other devices and power outlets, have your colleague press the button that's just above the large green circle on the computer. A three second ECG should be recorded by this action. You can touch any point on the graph to get the mV value and the time at that point. Use this technique to determine your ECG intervals and record the data in the table below.

**Standard resting ECG interval times (taken from LabQuest Vernier Lab instructions)**

| TABLE 4.1 | |
|---|---|
| **ECG INTERVAL** | **STANDARD RESTING ECG TIMES (SEC)** |
| P-R | 0.12 to 0.20 |
| QRS | Less than 0.12 |
| Q-T | 0.30 to 0.40 |

Read, and then decide on a hypothesis and how to run the experiment *before* you begin.

Choose the physical activity you feel comfortable performing in the lab (jog in place, take a single stair step up and down for 3-5 min and hold the rail, etc.). Measure your heart rate and then perform an ECG as described previously to establish a resting rate. Then, perform the activity for 5 minutes, and measure the HR and ECG again. Use individual's readings for each measurement, and if working with someone, then calculate the group average (minimum of three subjects) for each state for both HR and ECG.

Make a hypothesis of the outcome of your experiment. Will HR go up, go down, or stay the same? What do you think might happen to P-R interval, QRS, Q-T, and R-R?

1. Set up how you will do the experiment. Who will exercise? Who will keep time? Who will take EC? (The subject should be healthy with no heart problems.)
2. Take the resting HR and ECG and record in Table 4.2.
3. Perform exercise, retake the HR and ECG immediately afterwards, and record in Table 4.2.

**Student's ECG interval times and calculated heart rate before and after exercise**

| TABLE 4.2 | | |
|---|---|---|
| **YOUR ECG INTERVAL AND HEART RATE** | **RESTING** | **EXERCISE** |
| P-R | | |
| QRS | | |
| Q-T | | |
| R-R | | |
| Heart Rate | | |

Authored by James Hammarback

4. Review your hypothesis. Did you predict the outcome of the EKG difference between before and after exercise? Explain your reasoning for your hypothesis and for the outcome of the experiment.

5. Which intervals changed as a result of exercise? Which intervals (that you can see) did not change?

6. Which interval represents the signal to the ventricles prior to contracting?

   Which interval represents the signal to repolarize the ventricles?

7. What is tachycardia? What is bradycardia?

8. What are other key outcomes/diagnoses that an EKG can show us? List two. (Think of the more obvious issues with the heart, since it is an ECG, or respiratory/lung issues.)

## EXTENSION QUESTIONS

1.  How does the delay of the impulse at the atrioventricular node contribute to cardiac function?

2.  Why is the plateau phase so critical to cardiac muscle function?

*Be sure to get your completed work checked off by a member of the lab staff and then keep this hand out for your review.*

# 5

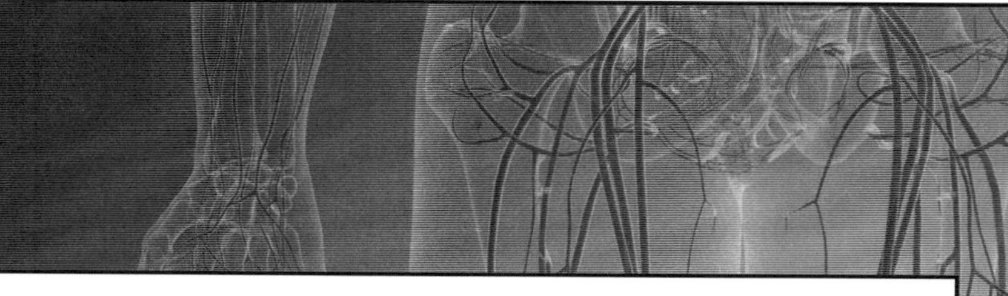

# CARDIOVASCULAR SYSTEM– BLOOD VESSELS
## PRE-LAB

Name: _________________________   Section: __________   Date: _________

## LEARNING OBJECTIVES

1. Compare and contrast the structure of arteries, veins, and capillaries.

2. Identify the layers or tunics of the artery and vein.

3. Understand the difference between heart rate and blood pressure.

4. Observe and measure the effects of exercise on heart rate and blood pressure, specifically systolic and diastolic pressures.

5. Identify the blood vessels of the cardiovascular system (see lab handout for list).

*Checklist to complete* **before entering** *the science skills lab (SSL):*

☐ Actively read this packet of information.

☐ Complete the charts, tables or labeling and answer questions using your own words.

☐ Complete the electronic digital pre-lab quiz on Bb by Sunday.

☐ Review the attached anatomy list and take it to lab with you. Jot down key descriptive identifying words that aid in your lab test preparations.

*You will need to have a* **microscope** *or access to a microscope in an A&P lab and a blood pressure cuff (sphygmomanometer) with stethoscope in order to complete this lab.*

From the heart, blood flows through the bigger **arteries,** which branch into more narrow **arterioles,** and then branch further still into the **capillaries,** which are small and thin enough to perform gas exchange at various body tissues. On the way from the tissues to the heart, capillaries join and widen to become **venules,** and then widen more to become **veins,** which **return blood to the heart.**

The **heart** is a complex muscle that consists of two pumps: one that pumps blood through **pulmonary circulation** to the lungs, and the other that pumps blood through **systemic circulation** to the rest of the body's tissues (and the heart itself). The heart is **asymmetrical,** with the left side being larger and stronger than the right side, so it can pump blood to the entire body.

The **right atrium** receives **deoxygenated** blood from the systemic circulation through the major veins: the **superior vena cava,** which drains blood from the head and from the veins that come from the arms, as well as the **inferior vena cava,** which drains blood from the veins that come from the lower organs and the legs. The right atrium also receives deoxygenated blood from the coronary circulation via the **coronary sinus.** This deoxygenated blood then passes to the **right ventricle** through the **tricuspid valve,** which prevents the backflow of blood. After it is filled, the right ventricle contracts, pumping the blood to the lungs for reoxygenation. The **left atrium** receives the **oxygen-rich** blood from the lungs. This blood passes through the **bicuspid valve** to the **left ventricle** where the blood is pumped into the **aorta.** The aorta is the major artery of the body, taking oxygenated blood to the organs and muscles of the body. This pattern of pumping is referred to as **double circulation and is found in all mammals.** Know this circulation for lab and class tests. **Use the models in lab and dissections to follow the flow of blood.**

## ACTIVITY 1: MICROSCOPE SLIDE—CROSS SECTION THROUGH HUMAN ARTERIES AND VEINS

*Read the content provided in your open resource text (i.e.: Lumen) and the following introductory paragraphs. Then, complete Activity 1 in lab.*

**Arteries** *take blood* **away from the heart.** *The main artery of the systemic circulation is the aorta; it branches into major arteries that take blood to different limbs and organs. The aorta and arteries near the heart have thick muscular but elastic walls that respond to and smooth out the pressure differences caused by the beating heart. Arteries farther away from the heart have more muscle tissue in their walls that can constrict to affect flow rates of blood.*

Arteries have a smaller lumen, a hollow passageway through which blood flows, than veins, a characteristic that helps to maintain the pressure of blood moving through the system. Together, their thicker walls and smaller diameters give arterial lumens a more rounded appearance in cross section than the lumens of veins [Also, when we cut veins in cross-section, they collapse since the walls are so thin.] The **endothelium** is the thin layer of cells that lines the interior surface of blood vessels, forming an interface between circulating blood in the lumen and the rest of the vessel wall. Endothelial cells reduce turbulence of the flow of blood, allowing the fluid to be pumped farther. Surrounding the endothelial layer are smooth muscle cells that can contract and relax to regulate the diameter of blood vessels. Find the three tunics/layers of each vessel wall.

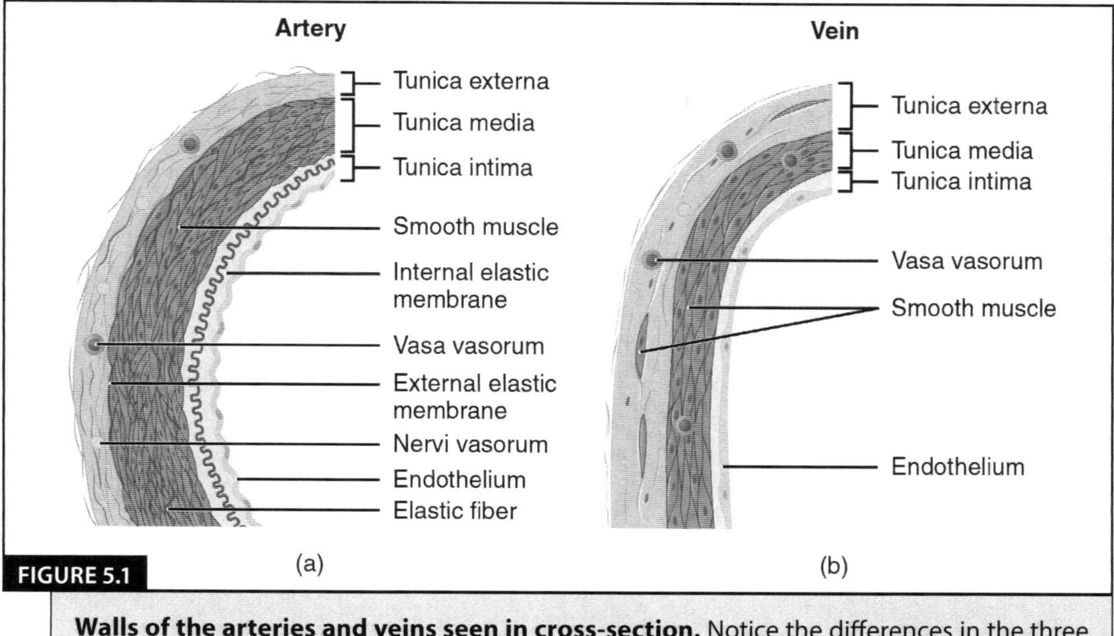


**FIGURE 5.1**

**Walls of the arteries and veins seen in cross-section.** Notice the differences in the three tunics/layers and what each is composed of in addition to elastic fibers/membrane.

OpenStax College [CC BY 3.0 (https://creativecommons.org/licenses/by/3.0)]

In order to deliver fresh blood to the peripheral organs, both heart action and the muscle in the wall of the arteries need to work hard. Unlike arteries, veins have less muscle action, and so they have thinner muscular walls (seen in microscopic cross-section); they are not supported by valves that prevent the backflow of the blood, but primarily by the contraction of the skeletal musculature that surrounds them. Thus, walking and exercising are beneficial for the heart health and help with circulation.

List structures that are unique to an artery or a vein, compared to each other, in the table.

| ARTERY | VEIN |
| --- | --- |
|  |  |

## ACTIVITY 2: MEASURING BLOOD PRESSURE

*Read the content provided in your open resource text (i.e.: Lumen) and the following introductory paragraphs. Then, complete Activity 2 in lab.*

The main purpose of the heart is to **pump blood** through the body; it does so in a repeating sequence called **the cardiac cycle.** The cardiac cycle is the flow of blood through the heart coordinated by electrochemical signals that cause **the heart muscle to contract and relax.** In each cardiac cycle, a sequence of **contractions** pushes out the blood, pumping it through the body; this is followed by a **relaxation** phase, where the heart fills with blood. These two phases are called the **systole** (contraction) and **diastole** (relaxation).

Blood pressure is measured in millimeters of mercury (mm Hg). A typical blood pressure is 120/80 mm Hg, or "120 over 80." The first number represents the pressure when the heart contracts and is called the systolic blood pressure. The second number represents the pressure when the heart relaxes and is called the diastolic blood pressure. Blood pressure is in the normal range (defined by the American Heart Association) between 90–119 mmHg for systolic pressure and a diastolic reading of between 60–79 mmHg. The pulse pressure is the difference between the systolic and diastolic pressures or the change in pressure with each heartbeat.

$$\mathbf{P}_{pulse} = \mathbf{P}_{systolic} - \mathbf{P}_{diastolic}$$

As arteries become less compliant with age, the pulse pressure usually increases. As the pulse pressure increases with age, this adds to the risk of heart disease.

Another measure of cardiovascular health is the mean arterial pressure (MAP), which is the average pressure in the arteries over one cardiac cycle. It is calculated as:

$$\mathbf{MAP} = \mathbf{P}_{diastolic} + 1/3\mathbf{P}_{pulse} \quad \text{OR} \quad \mathbf{MAP} = \mathbf{P}_{diastolic} + (\mathbf{P}_{systolic} - \mathbf{P}_{diastolic})/3$$

MAP is normally between 65 and 110 mmHg (with 70mmHg enough to sustain organ activity). If the MAP falls below this number for a substantial amount of time, vital organs will not get enough oxygen perfusion and will become **hypoxic,** a condition called **ischemia.** So MAP is a good indicator of perfusion pressure, thus showing the health care worker whether the cardiovascular system is able to sustain the organs.

- Why do you think high blood pressure over time leads to detrimental impacts on health?

In lab, **review** the steps to measuring blood pressure and pulse/heart rate. Read below.

**Materials: Sphygmomanometer** (blood pressure cuff)

1. Deflate the air bladder of the cuff and place it around the upper arm so it fits snugly. If you are right-handed, you should hold the bulb/pump in your left hand to inflate the cuff. Hold it in the palm so your fingers can easily reach the valve at the top to open and close the outlet to the air bladder.

2. Put the head of the stethoscope just under the edge of the cuff, a little above the crease of the person's elbow.

3. Inflate the cuff with brisk squeezes of the bulb. Watch the pressure gauge as you do it; you should go to around 150 mmHg or until the pulse is no longer heard. At this point, blood flow in the underlying blood vessel is cut off by pressure in the cuff.

4. At around 150, slightly open the valve on the air pump (held in your left hand). This part takes practice; it is important that you don't let the air out too suddenly.

5. Now, pay attention to what you hear through the stethoscope as the needle on the pressure gauge falls. You will be listening for a slight "blrrp" or something that sounds like a "prrpshh." The first time you hear this sound, note the reading on the gauge. This value is the **systolic blood pressure.**

6. The sounds should continue and become louder in intensity. Note the reading when you hear the sound for the last time. This is the **diastolic blood pressure.**

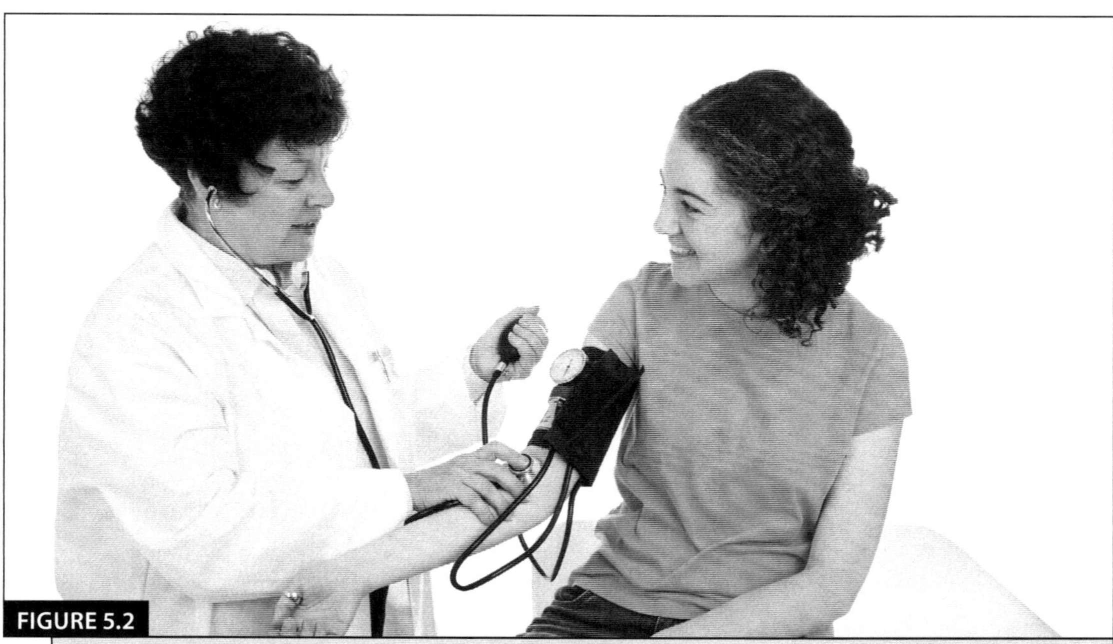

**FIGURE 5.2**

**Blood pressure cuff or sphygmomanometer and the brachial artery in the arm**

## MEASURING HEARTBEAT USING A STETHOSCOPE

Auscultation of the heart means to listen to and study the various sounds arising from the heart as it pumps blood. These sounds are the result of vibrations produced when the heart valves close, and blood rebounds against the ventricular walls or blood vessels. The heart sounds may be heard by placing the ear against the chest or by using a stethoscope. Two major sounds can be heard.

- **First heart sound:** Produced at the beginning of systole when the atrioventricular (AV) valves close and the semilunar (SL; the aortic and pulmonary) valves open. This sound has a low-pitched tone commonly termed the "lub" sound of the heartbeat.

- **Second heart sound:** Occurs during the end of systole and is produced by the closure of the SL valves, the opening of the AV valves, and the resulting vibrations in the arteries and ventricles. Due to the higher blood pressures in the arteries, the sound produces a higher pitch than the first heart sound. It is commonly referred to as the "dup" sound.

**Note:** In lab, measure your heart rate and the heart rate of two test subjects over a 30-second period. Multiply your heart rates by two to get your total heart rate in a minute.

## MEASURING PULSE

You should know that your "pulse" refers both to the physical thump created in your arteries by the contraction of your heart muscles and the number of these thumps your heart causes per minute. You have seven pulse points—places where arteries come close to your skin—on your body.

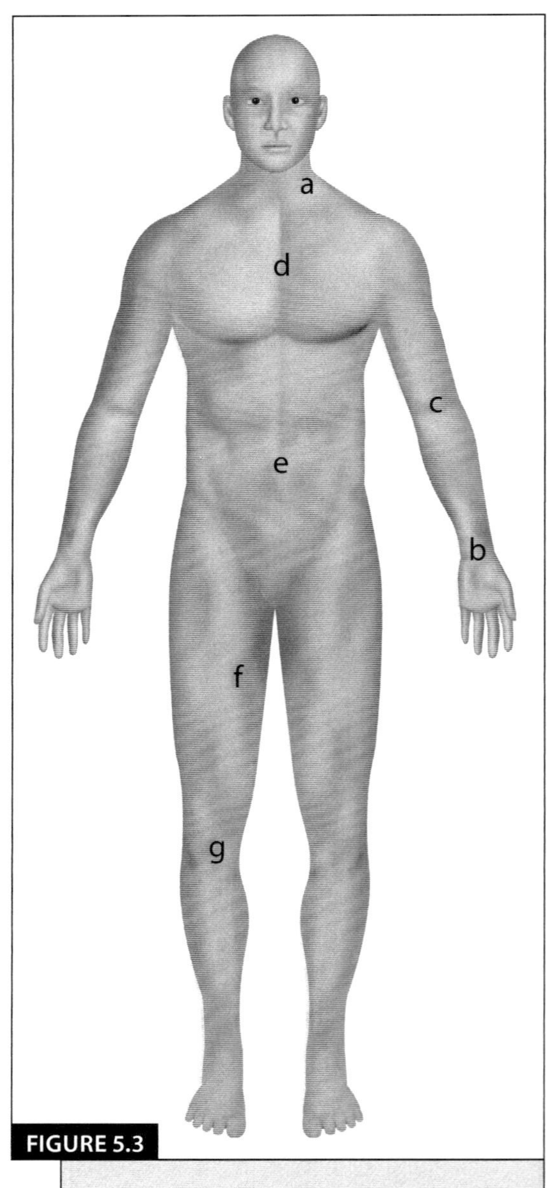


**FIGURE 5.3**

- **a.** carotid arteries (located on your neck)
- **b.** radial arteries (on your wrists)
- **c.** brachial arteries (on your arms)
- **d.** aortic arch (by your heart)
- **e.** abdominal aorta (near your stomach)
- **f.** femoral arteries (on your thighs)
- **g.** popliteal arteries (near your knees)

## MEASURE USING YOUR RADIAL ARTERY

To find your radial artery (the most common point from which people take pulses), hold one hand straight out, elbow bent, palm relaxed and facing up. Raise your thumb slightly skyward, as if holding an apple or a tennis ball, to create a small pocket under your thumb at the top of your wrist where you will place the tips of your index and middle finger. (Do not use your thumb, which also has a pulse and could cause counting confusion.)

- Count the beats for 30 seconds and multiply by two. This is your pulse rate: ______

## MEASURE USING YOUR CAROTID ARTERY

Neck pulse points are stronger and more accessible. The carotid is located just below your jaw in the groove where your head and neck meet, on either side of your windpipe. Use your index and middle fingertips to feel around in the groove for a pulsation.

## ACTIVITY 3: IDENTIFY THE BLOOD VESSELS

Identify the blood vessels above as well as others on the anatomy list on the lab activities handout on models in lab and/or on dissections

**Arteries:** aorta and its branches, subclavian arteries and branches, arteries of the arm, arteries to the head including the Circle of Willis, descending aorta and branches, and branches of the external iliac arteries (femoral artery, popliteal artery, anterior tibial artery, plantar arch)

**Veins:** veins of the upper appendages to the subclavian veins, brachiocephalic veins and superior vena cava, veins of the head to the internal jugular veins, veins of the thoracic cage, abdominal veins and the hepatic portal vein, veins of the lower appendages to the saphenous veins, iliac veins, and inferior vena cava.

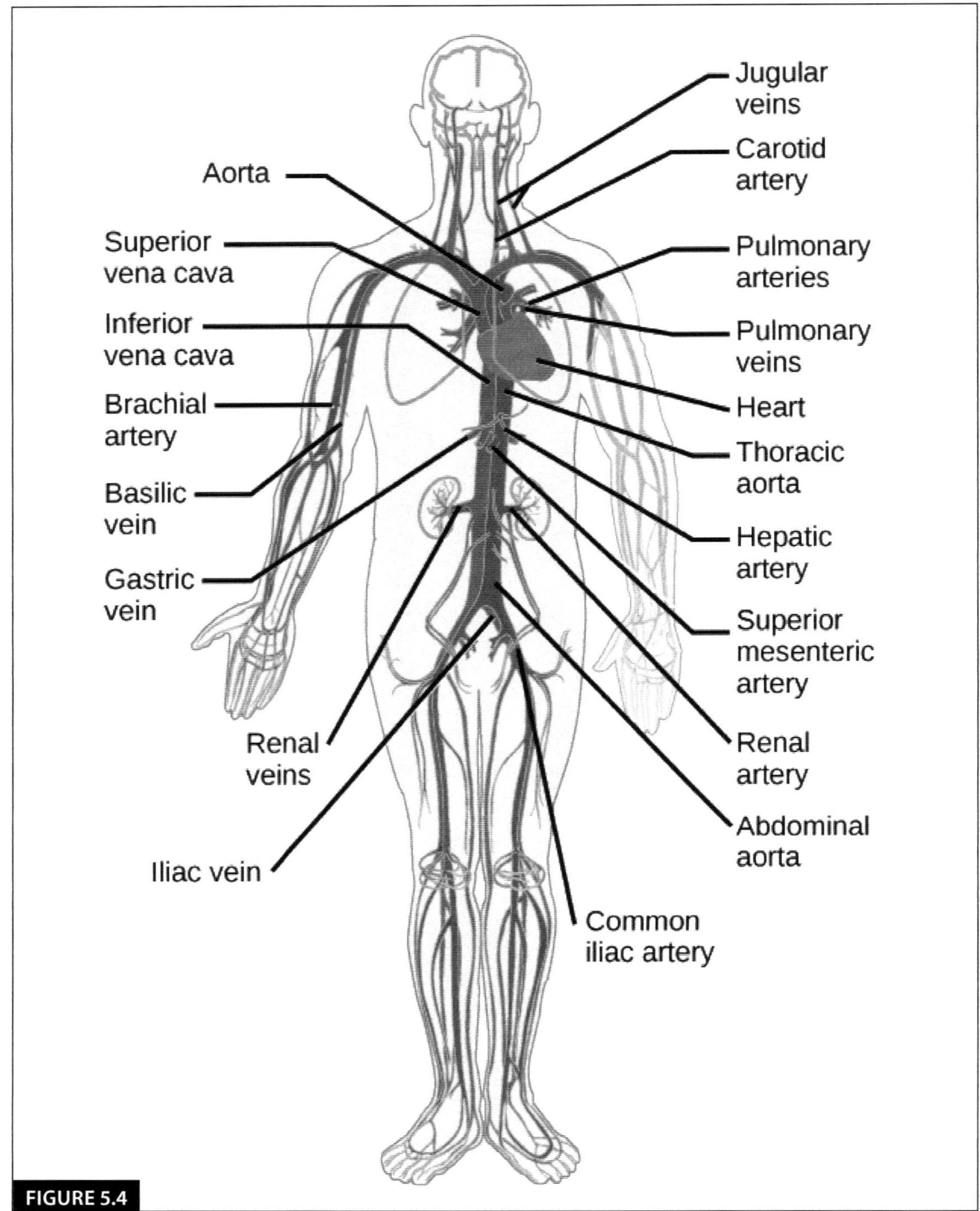


**FIGURE 5.4**

**Major arteries and veins.** The blood from the heart is carried through the body by a complex network of blood vessels. This diagram illustrates the major human arteries and veins of the human body. Mammalian Heart and Blood Vessels [https://cnx.org/contents/GFy_h8cu@11.10:ZdC2EWuz@6/Mammalian-Heart-and-Blood-Vessels] by OpenStax (CC BY 4.0)

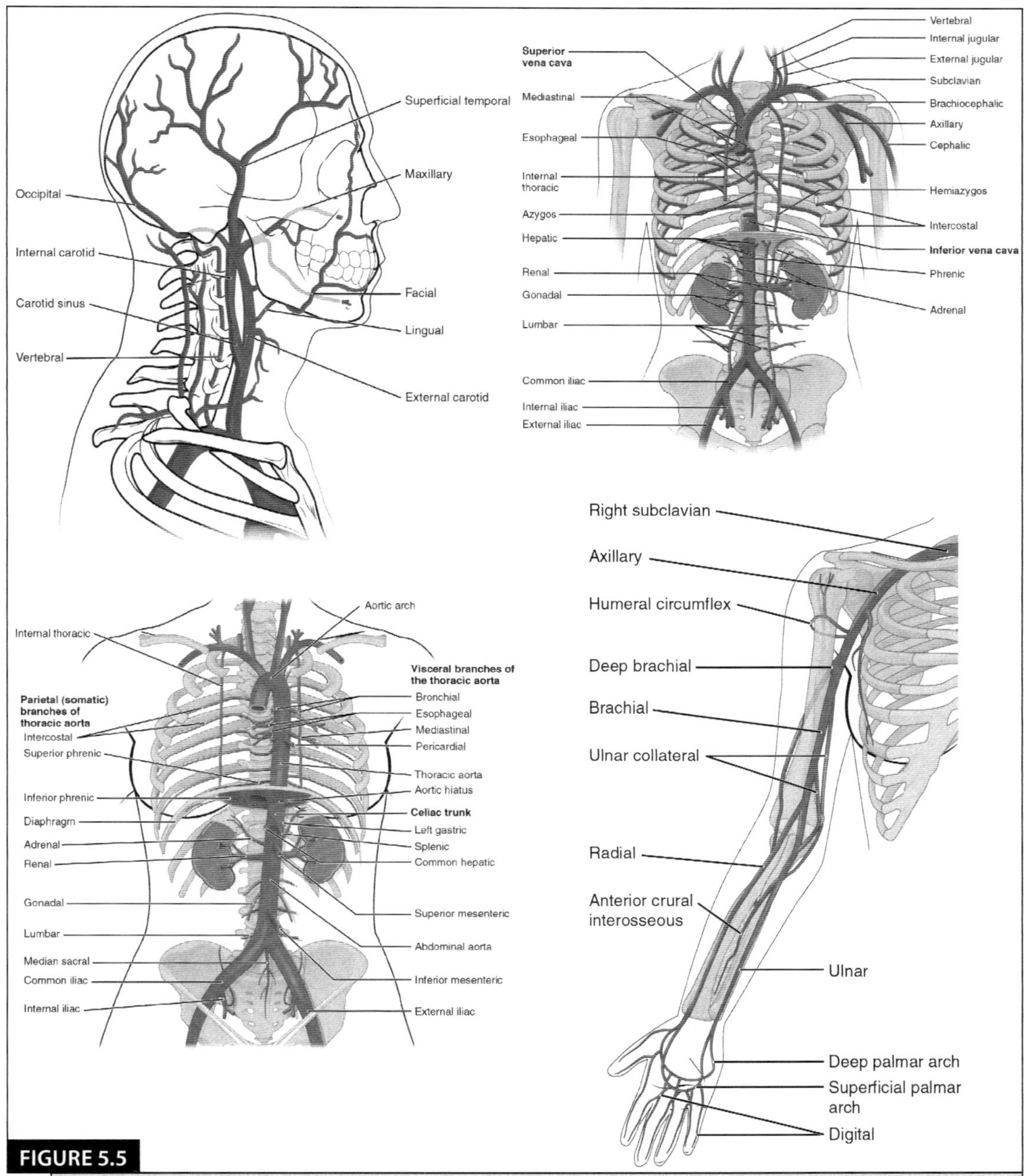


**FIGURE 5.5**

### Blood vessels of the human body–head, torso, and arm.

OpenStax College, Mammalian Heart and Blood Vessels. October 17, 2013. CC BY Creative Commons Attribution 4.0 International License

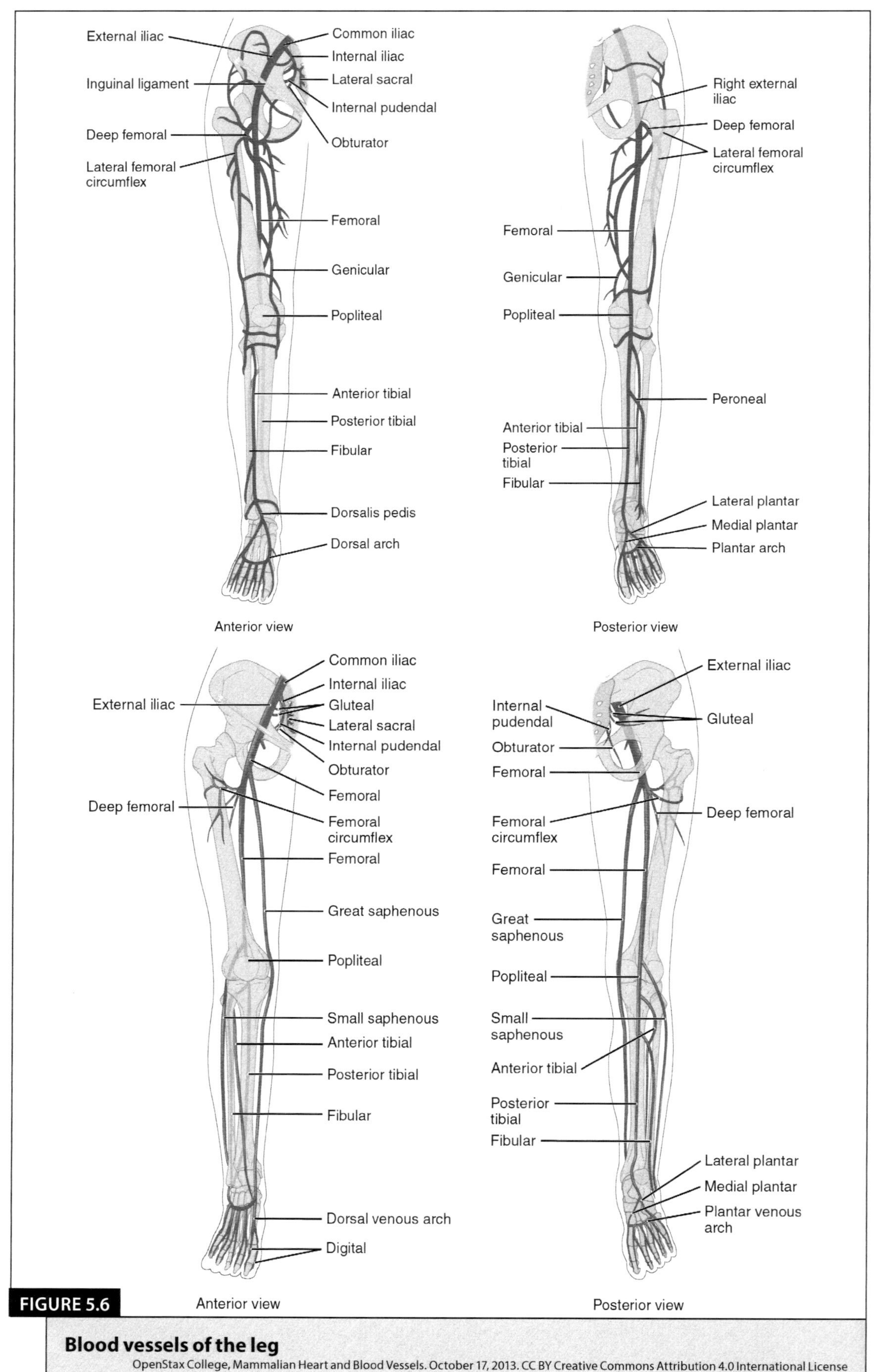


**FIGURE 5.6**

**Blood vessels of the leg**

OpenStax College, Mammalian Heart and Blood Vessels. October 17, 2013. CC BY Creative Commons Attribution 4.0 International License

# 5

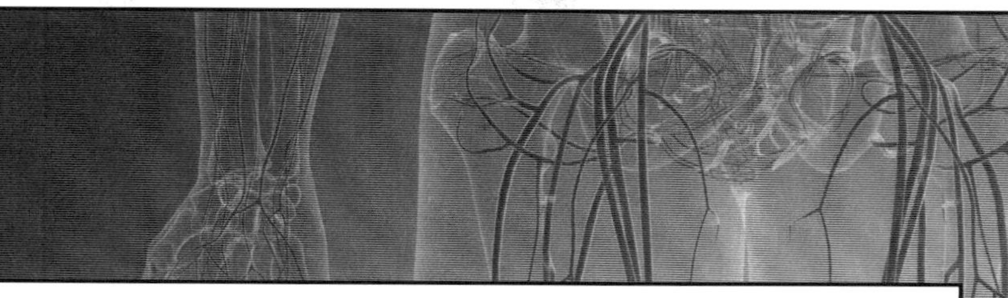

# CARDIOVASCULAR SYSTEM–
# BLOOD VESSELS
## IN-LAB ACTIVITIES

Name: _______________________________    Section: ____________    Date: __________

## LEARNING OBJECTIVES

1. Compare and contrast the structure of arteries, veins, and capillaries.

2. Identify the layers or tunics of the artery and vein.

3. Understand the difference between heart rate and blood pressure.

4. Observe and measure the effects of exercise on heart rate and blood pressure, specifically systolic and diastolic pressures.

5. Identify the blood vessels of the cardiovascular system. (*You will spend much of the* ***2–3 hours*** *of your lab time on this section, so do not spend too much time on the other two activities.*)

*You will need to have a* **microscope** *or access to a microscope in an A&P lab and a blood pressure cuff (sphygmomanometer) with stethoscope in order to complete this lab.*

## ACTIVITY 1: OBSERVE THE MICROSCOPE SLIDE OF THE ARTERY AND VEIN

Observe the slide in lab and make a **drawing** of both human artery and vein cross-section; **label** the layers or tunicas of each vessel.

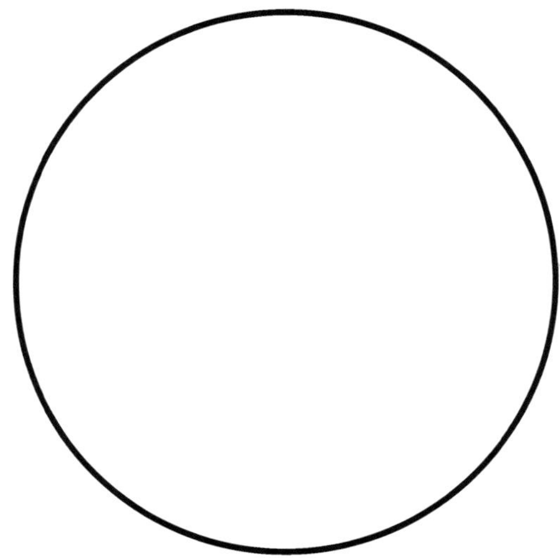

1.  What is the difference between arteries and veins that you can easily observe and can guide you in distinguishing between them?

## ACTIVITY 2: EFFECT OF EXERCISE ON BLOOD PRESSURE

1.  Write the formula to calculate mean arterial pressure (MAP).

2.  Before you begin setting up your experiment, write a brief introduction that includes your hypothesis of the outcome below (The introduction should explain why you would do the experiment).

---

---

---

---

*Read, then decide on a hypothesis and how to run the experiment* **before** *you begin.*

Choose the physical activity you feel comfortable performing in the lab (jog in place, take a single stair step up and down for 3–5 min and hold the rail, etc.). Measure your heart rate and blood pressure in **resting state** (Do you think the subject should be sitting or standing?), *then* perform the **activity for 5 minutes** and measure the HR and blood pressure again. Use individual's readings for each measurement, and then calculate the group average (minimum of three subjects) for each state for both HR and BP.

1. Make a hypothesis of the outcome of your experiment. Will BP go up, go down, or stay the same? Will both systolic and diastolic pressure rise? What will happen to the MAP average?

---

---

2. Set up how you will do the experiment. Who will exercise? Who will keep time? Who will take BP? (The subject should be healthy with no heart problems).

---

---

3. Take the resting HR and BP.

4. Perform exercise and retake the HR and BP immediately afterwards.

**Resting Heart Rate Compared to HR After Exercise**

| TABLE 5.1 | RESTING HEART RATE/PULSE: | AFTER ACTIVITY HEART RATE/PULSE |
|---|---|---|
| 1st subject | | |
| 2nd subject | | |
| 3rd subject | | |
| Average | | |

**Resting Blood Pressure Compared to After Exercise**

| TABLE 5.2 | RESTING SYSTOLIC PRESSURE | RESTING DIASTOLIC PRESSURE | AFTER ACTIVITY SYSTOLIC PRESSURE | AFTER ACTIVITY DIASTOLIC PRESSURE |
|---|---|---|---|---|
| 1st subject | | | | |
| 2nd subject | | | | |
| 3rd subject | | | | |
| Average | | | | |

## DISCUSSION QUESTIONS

**1.** Did your Blood Pressure change with activity? _________________________ Why?

_________________________________________________________________

_________________________________________________________________

_________________________________________________________________

_________________________________________________________________

**2.** Was there a different outcome than you hypothesized? _______________ Explain.

_________________________________________________________________

_________________________________________________________________

_________________________________________________________________

**3.** Compare the systolic pressure to the diastolic pressure.

    **a.** **How** was the diastolic different after exercise?

_________________________________________________________________

_________________________________________________________________

**b. Why** was the systolic different?

_______________________________________________

_______________________________________________

4. Should you take resting values for heart rate and blood pressure sitting or standing? Why?

_______________________________________________

_______________________________________________

_______________________________________________

_______________________________________________

5. What equipment is used to measure heart rate? _______________________

   To measure blood pressure? _______________________

6. Why should you **not** use your thumb to measure pulse?

_______________________________________________

_______________________________________________

_______________________________________________

7. Why is it better to get the pulse pressure and MAP when dealing with heart patients? (You may have to research this answer.)

_______________________________________________

_______________________________________________

_______________________________________________

_______________________________________________

8. Look at the differences between capillary walls (junctions versus spaces), and define the following terms. You should include where each type of capillary can be found and the usefulness of its structure.

    **a.** tight junction

    **b.** intercellular cleft

    **c.** fenestration

    **d.** sinusoid

# ACTIVITY 3: IDENTIFY ON THE CIRCULATORY BOARDS AND OTHER MODELS IN LAB THE FOLLOWING BLOOD VESSELS —

*Pictures and legends will be available to you in lab.*

1. Identify the following microscopic structures on slides of **arteries and veins:**
    **a.** Artery
        **i.** tunica interna (tunica intima)
            1. endothelium
            2. internal elastic membrane
        **ii.** tunica media
            1. smooth muscle fibers
            2. external elastic membrane
        **iii.** tunica externa (tunica adventitia)

  **b.** Vein

     **i.** tunica interna (tunica intima)

       1. endothelium

     **ii.** tunica media

       1. smooth muscle fibers

       2. elastic fibers

     **iii.** tunica externa (tunica adventitia)

**2.** Identify the following **arteries** associated with the aortic arch on various models, making sure to identify left and right where appropriate:

  **a.** aorta

     **i.** ascending aorta

     **ii.** aortic arch

       1. brachiocephalic artery (innominate artery)

         a. right common carotid artery

         b. right subclavian artery

           i. right internal thoracic artery (internal mammary artery) (on the head/bust muscular lab model)

           ii. right vertebral artery

       2. left common carotid artery

       3. left subclavian artery

         a. left internal thoracic artery (only in pictures)

         b. left vertebral artery

**3.** Identify the following **arteries** associated with the upper extremities on various models, making sure to identify left and right where appropriate:

  **a.** axillary artery

  **b.** brachial artery

  **c.** radial artery

  **d.** ulnar artery

  **e.** palmar arches

  **f.** digital arteries

4. Identify the following **arteries** associated with the blood supply to the brain on various models, making sure to identify left and right where appropriate:

   a. common carotid artery

      i. carotid sinus

      ii. external carotid artery

         1. facial artery

         2. superficial temporal artery

      iii. internal carotid artery

         1. carotid canal (temporal bone in the base of the skull)

      iv. vertebral artery

      v. cerebral arterial circle (Circle of Willis)

         1. basilar artery

         2. posterior cerebral arteries

         3. posterior communicating artery

         4. middle cerebral arteries

         5. anterior communicating artery (difficult to see on models)

         6. anterior cerebral arteries

5. Identify the following **arteries** associated with the descending aorta on various models, making sure to identify left and right where appropriate:

   a. descending thoracic aorta

      i. intercostal arteries

   b. aortic hiatus in the diaphragm

   c. descending abdominal aorta

      i. inferior phrenic artery

      ii. celiac trunk (celiac artery)

         1. common hepatic artery

            a. hepatic arteries

            b. cystic artery

         2. splenic artery

         3. left gastric artery

      iii. superior mesenteric artery

      iv. renal artery

      v. gonadal artery (testicular artery; ovarian artery)

      vi. inferior mesenteric artery

6. Identify the following **arteries** associated with the pelvic region and lower extremities on various models, making sure to identify left and right where appropriate:
   a. common iliac artery
      i. internal iliac artery
      ii. external iliac artery
         1. femoral artery
         2. popliteal artery
         3. anterior tibial artery
            a. dorsalis pedis artery
            b. arcuate arch
         4. posterior tibial artery
         5. fibular artery (peroneal artery)
         6. plantar arch

7. Identify the following **veins** associated with the upper extremities on various models, making sure to identify left and right where appropriate:
   a. digital veins
   b. radial vein
   c. ulnar vein
   d. cephalic vein
   e. basilic vein
   f. brachial vein
   g. axillary vein
   h. subclavian vein
   i. brachiocephalic vein
   j. superior vena cava

8. Identify the following **veins** associated with the head, neck, and thorax on various models, making sure to identify left and right where appropriate:

   a. superior sagittal sinus

   b. inferior sagittal sinus

   c. transverse sinus

   d. internal jugular veins

       i. jugular foramen (in the border of the temporal and occipital bones in the base of the skull)

   e. facial vein

   f. external jugular vein

   g. azygos vein—can be found on hearts' posterior SVC and on large "Maximan" in Allman

   h. intercostal veins

9. Identify the following **veins** associated with the lower extremities on various models, making sure to identify left and right where appropriate:

   a. small saphenous vein

   b. great saphenous vein

   c. anterior tibial vein

   d. posterior tibial vein

   e. femoral vein

   f. external iliac vein

   g. internal iliac vein

   h. common iliac vein

   i. inferior vena cava

10. Identify the following **veins** associated with the abdomen on various models, making sure to identify left and right where appropriate:

    a. renal veins

    b. superior mesenteric vein

    c. inferior mesenteric vein

    d. splenic vein

    e. hepatic portal vein (light blue or purple on the liver model)

    f. hepatic veins (blue on the liver model)

*Be sure to get your completed work checked off by a member of the lab staff and then keep this hand out for your review.*

# IMMUNE AND MICROBIOLOGY LAB
## PRE-LAB

Name: _____________________________  Section: __________  Date: _________

*The activities in this lab are associated with the lecture materials on the Immune System.*

## OBJECTIVE

In this lab you will learn about the immune system and perform laboratory activities that demonstrate effects of disinfectants and the importance of antibodies and about the use of the innate vs adaptive immune system.

## INSTRUCTIONS FOR STUDENTS

Read the entire laboratory procedure and complete all pre-lab activities before coming to lab. You must print off this pre-lab and complete by hand. There are computer lab resources and Learning Centers for limited daily free printing. Once you come to lab, follow the steps provided for each exercise, take good notes and make sure to get checked off by the lab instructor once you finish.

## LEARNING OBJECTIVES

After completing this laboratory experiment, you should be able to do the following:

1. Students will be able to describe the role of the immune system.

2. Students will take an environmental sample to grow bacteria on agar plates. This can demonstrate that the innate first line of defense protects us from the diversity of bacteria and germs around us.

*Checklist to complete* **before entering** *the science skills lab (SSL):*

☐ Actively read this packet of information.

☐ Complete the charts, tables or labelling and answer questions using your own words.

☐ Complete the electronic digital pre-lab quiz on Bb by Sunday.

☐ Review the attached anatomy list and take to lab with you, in order to jot down key descriptive identifying words that aid in your lab test preparations.

## IMMUNE SYSTEM INTRODUCTION AND BACKGROUND

The immune is divided into two components: the **innate immune system,** which is nonspecific toward a particular kind of pathogen, and the **adaptive immune system,** which is specific and has memory. Innate immunity is the first line of defense against diseases and depends on physical and chemical barriers that work on all pathogens. The innate immune system can also release chemical signals that produce **inflammation** and fever responses as well as mobilizing other immune cells. The adaptive immune system mounts a highly specific response to substances and organisms that do not belong in the body. The adaptive system takes longer to respond and has a memory system that allows it to respond with greater intensity should the body reencounter a pathogen even years later.

## EXTERNAL AND CHEMICAL BARRIERS

The body has significant physical barriers to potential pathogens. The skin contains the protein keratin, which resists physical entry into cells. Other body surfaces, particularly those associated with body openings, are protected by the mucous membranes. The sticky mucus provides a physical trap for pathogens, preventing their movement deeper into the body. The openings of the body, such as the nose and ears, are protected by hairs that catch pathogens, and the mucous membranes of the upper respiratory tract have cilia that constantly move pathogens trapped in the mucus coat up to the mouth. Many surfaces of the body have chemical environments that are hostile to microorganisms, for example the acidity of the skin and the stomach.

## INTERNAL DEFENSES

When pathogens enter the body, the innate immune system responds with a variety of internal defenses. These include the inflammatory response, phagocytosis, natural killer cells, and the complement system. White blood cells in the blood and lymph recognize pathogens as foreign to the body. For example, a **monocyte** is a type of white blood cell that circulates in the blood and lymph and develops into a **macrophage** after it moves into infected tissue where it may engulf foreign particles and pathogens. Another type of innate immune cell are the **Mast cells** which release chemicals in response to physical injury and play an important role in the allergic response by releasing histamines. The immune system has a number of chemical messengers that it uses to communicate with cells in different parts of the body. The cytokines are one such messenger that can influence cell migration and behaviors. For example, the **interferons** are cytokines that are released by infected cells that stimulate nearby cells to release antiviral molecules and engulf the damaged cell.

## THE INFLAMMATORY RESPONSE AND PHAGOCYTOSIS

The first cytokines to be produced upon infection will stimulate **inflammation,** which is a localized redness, swelling, heat, and pain. Inflammation is a response to physical trauma, such as a cut or a blow, chemical irritation, and infection by pathogens (viruses, bacteria, or fungi). The types of white blood cells that arrive at an inflamed site depend on the nature of the injury or infecting pathogen. A **neutrophil** is an early arriving white blood cell that engulfs and digests pathogens in a process called **phagocytosis.** Macrophages follow neutrophils and take over the phagocytosis function and are involved in the resolution of an inflamed site, cleaning up cell debris and pathogens. A **lymphocyte** is a white blood cell that contains a large nucleus. Most lymphocytes are associated with the adaptive immune response, but infected cells are identified and destroyed by natural killer cells, the only lymphocytes of the innate immune system. **Natural killer (NK) cell** are lymphocytes that can kill cells infected with viruses (or cancerous cells). NK cells identify intracellular infections, especially from viruses, and induces programmed cell death (**apoptosis**) and the resulting cell debris are engulfed by phagocytes. Figure 6.1 outlines some parts of the innate immune response that may be involved in a cut.

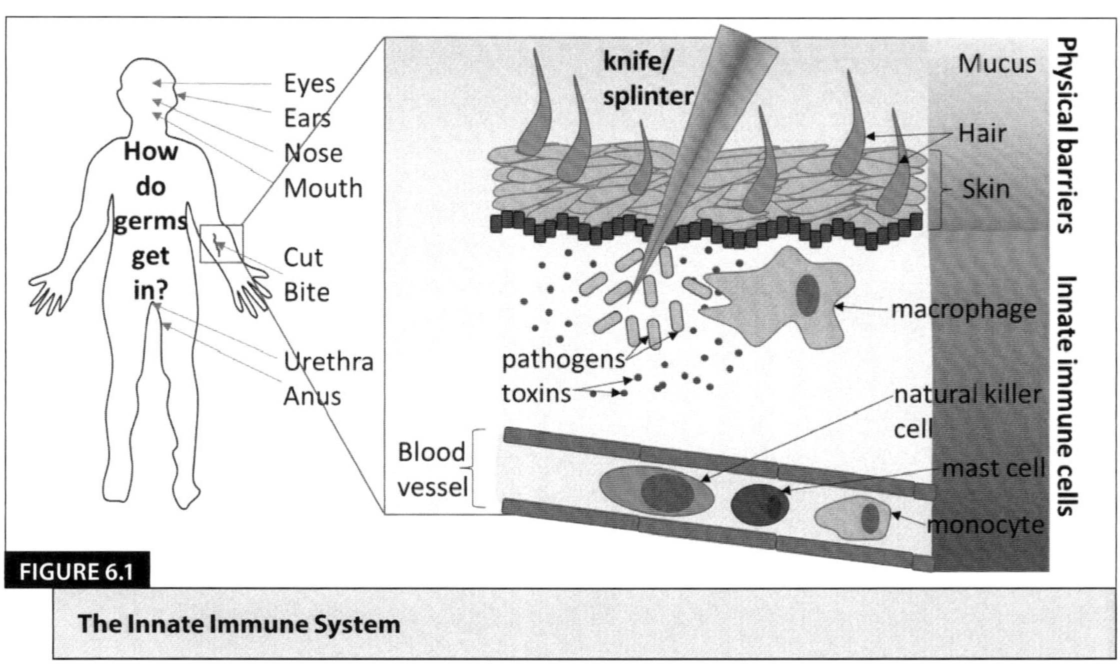


**FIGURE 6.1**

**The Innate Immune System**

# ADAPTIVE IMMUNITY

The adaptive, or acquired, immune response takes days or even weeks to become established, however, **adaptive immunity** is more specific to an invading pathogen and possesses memory. **Adaptive immunity** is an immunity that occurs after exposure to an antigen either from a pathogen or a vaccination. An **antigen** is a molecule that stimulates a response in the immune system. This part of the immune system is activated when the innate immune response is insufficient to control an infection. There are two types of adaptive responses: the **cell-mediated immune response,** which is controlled by activated **T cells,** and the **humoral immune response,** which is controlled by activated **B cells** and antibodies. Adaptive immune cells can kill pathogens directly (T cells) or they can secrete antibodies (B cells) that enhance the phagocytosis of pathogens and disrupt the infection. Adaptive immunity also involves a memory to give the host long-term protection from reinfection with the same type of pathogen; on re-exposure, this host memory will facilitate a rapid and powerful response.

# HUMORAL IMMUNE RESPONSE

We looked at blood cell types and the antibodies with them in the Blood Lab. We should learn a little bit more about their role in immunity. As mentioned, an antigen is a molecule that stimulates a response in the immune system. Not every molecule is antigenic. Lymphocytes called B-cells participate in a chemical response to antigens present in the body by producing specific antibodies that circulate throughout the body and bind with the antigen when encountered. This is known as the humoral immune response. Each B cell has only one kind of antigen receptor that recognizes only one antigen, which makes every B cell different. When a B cell encounters the antigen that binds to its receptor, the antigen molecule is brought into the cell by endocytosis and

reappears on the surface of the cell bound to an **MHC class II molecule** and this B-cell may become activated to divide and make thousands of identical (clonal) cells. These daughter cells become either plasma cells, which secrete antibodies, or memory B-cells, which are quickly activated if the same infection occurs later. An **antibody,** also known as an immunoglobulin (Ig), that circulates in the blood stream and lymphatic system and can bind to its target antigen and lead to its destruction in a number of ways.

The production of antibodies by plasma cells in response to an antigen is called **active immunity** and describes the host's active response of the immune system to an infection or to a vaccination. There is also a **passive immune** response where antibodies come from an outside source, instead of the individual's own plasma cells, and are introduced into the host. For example, antibodies circulating in a pregnant woman's body move across the placenta into the developing fetus. The child benefits from the presence of these antibodies for up to several months after birth. In addition, a passive immune response is possible by injecting antibodies into an individual in the form of an anti-venom to a snake-bite toxin or antibodies in blood serum to help fight a hepatitis infection. More recent applications of passive immunity include fighting inflammatory diseases like rheumatoid arthritis and some antibodies are even being tested to treat cancer.

Answer the following questions:

1.  Compare and contrast innate immunity with adaptive immunity.

2.  What is the inflammatory response and what is its purpose for the body?

3.  If you were given a vaccine, what type of immunity would that be?

4.  If you were given shots of antibodies against rabies, what type of immunity would you be getting?

5.  Compare and contrast cell-mediated immunity with humoral immunity.

**6.** What are MHC proteins and why are they critical for self-recognition?

**7.** How can snake handlers become immune to the toxin secreted by the snakes they handle?

## ACTIVITY 1: MICRO LIFE AROUND US

Microorganisms, like other life forms, must have adequate nutrients and a favorable environment for growth. You do not see the bacteria around us because the environment may not favor their growth (but if you have ever seen the pink stain appear in the bath tub or shower, then you've seen them because they have had a good environment to grow in). In lab you will sample an environmental location of your choice and allow them to grow with optimal food, so you can see the variety of organisms we have around us.

**Preview the instructions below,** so that you will be better prepared to begin the experiment in lab.

A very simple experiment that is easily done in lab is the plating of microbes from our environment.

### MATERIALS

- Agar plates
- Sterile swabs
- Parafilm to seal the plates
- Sharpie Permanent pens
- Gloves

## METHODS

Put on your gloves. Take the plate of nutrient agar and the sealed sterile swab. Find an area that you would like to test for microbial contamination. With gloved hands open the sterile swab package, so that you can remove the swab without touching the swab itself. Remove the swab. Wet the swab with distilled or deionized water. With the other hand hold the plate with the agar side and remove the lid (only do this at the site you are testing)—try to hold that with the fingers of the same hand. Use the swab to swipe the area you want to test–you can move the swab back and forth over the area 2–4 times. Then **_very carefully_** lightly touch the surface of the agar without pushing into the gel. You will roll the swab gently over the surface of the agar so that your swipes cover most of the surface (see image below).

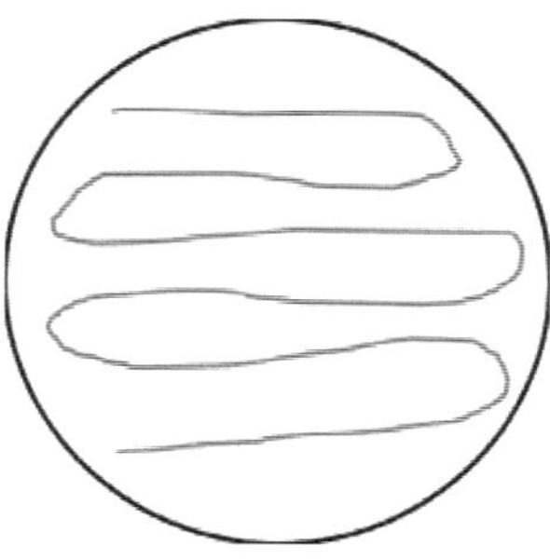

Then replace the cover of the agar plate, without further contaminating the plate. Replace the swab into the package then dispose of the swab in a proper trash receptacle. Once back in the lab, get a strip of parafilm or red tape that will go around the rim of the plate to seal it from further air, etc. Seal the plate around the sides (not over the top). (See below) Use the Sharpie pen to label your plate (label the plate on the edges, so you can see the microbes) by the **location** from where you got your sample, **name and date.** Place the plates upside down in the lab room where they will not be disturbed until the next lab period (usually a 1 week incubation time is good in room temperature).

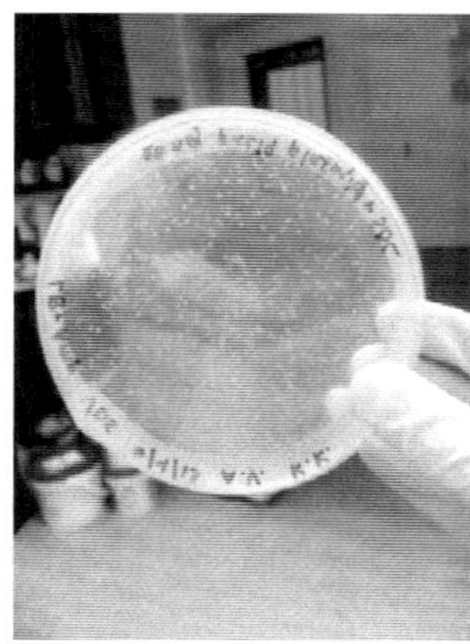

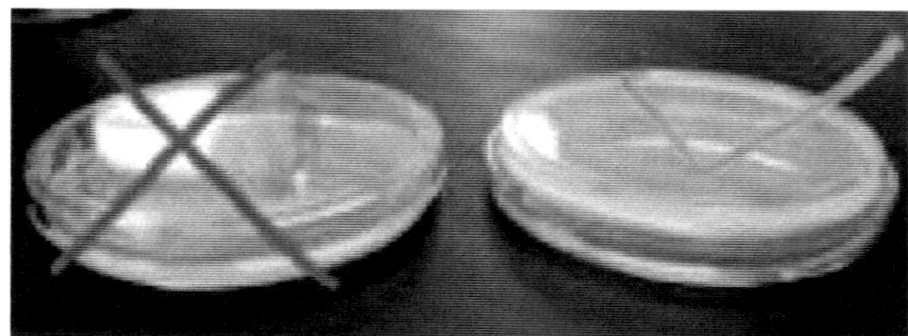

The next lab get your plate (with your labeled sample area on it–other students may have gone to the same area, so be sure it's yours). ***Do not*** unseal the plate. Observe the various bacterial groups on the plate. These round or odd shaped groups represent 1000's of bacteria, and we call each of these a colony. If they have different colors or edging or a surface sheen, then they are different types of microbes.

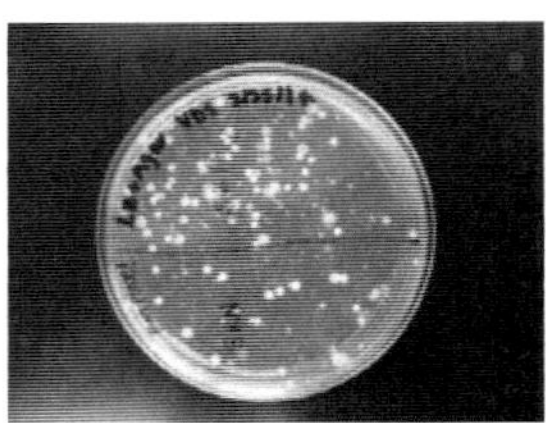

1.  How many different types of microbes do you have from your sample area?

    _______________________________________________________________________

2.  Compare these to the other plates in the lab group, and to the plates of known microorganisms.

    a.  Which areas/locations had the most microbes or the most variations of microbes on the plate?

        ___________________________________________________________________

    b.  Which areas/locations had the least?

        ___________________________________________________________________

    c.  Which areas are you surprised about, either with many colonies or with few colonies on them?

        ___________________________________________________________________

    d.  What other characteristics of the colonies did you notice?

        ___________________________________________________________________

        - If you have a colony that looks very "fuzzy" then that is a fungus.

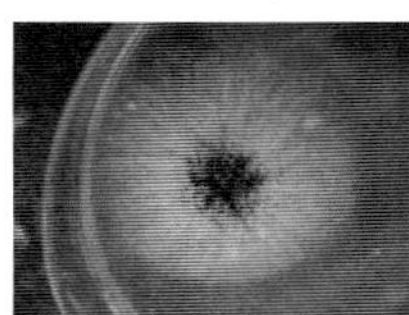

**3.** Place the plate into the RED biohazard bag for autoclaving (or bleaching) and disposal. Why does it need to go into the bag, everything is from our environment?

# 6

# IMMUNE AND MICROBIOLOGY LAB
## IN-LAB ACTIVITIES

Name: _________________________  Section: __________  Date: _________

## LEARNING OBJECTIVES

1. Learn and review the anatomy of the lymphatic or immune system.

2. Make streak plates and find out what common bacteria is around us. Use disinfectants or antibody discs on streak plates to determine which work best against common bacteria.

## ACTIVITY 1: LEARN THE ANATOMY OF THE LYMPHATIC OR IMMUNE SYSTEM

*Identify the following structures associated with the lymphatic system on models in the lab or pictures:*

1. lymph nodes

    a. inguinal nodes

    b. cervical nodes

    c. axillary nodes

2. lymph vessels (on microscope slide)

3. cisterna chili (on picture)

4. thoracic duct (left lymphatic duct)

5. right lymphatic duct

6. thymus gland

7. spleen

8. vermiform appendix
9. tonsils (find on the sagittal head in lab)
   a. pharyngeal tonsils
   b. palatine tonsils
   c. lingual tonsils

Answer the following questions.

1. What is the function of the lymph nodes?

   ______________________________________________________________

2. What is the function of the lymphocytes?

   ______________________________________________________________

3. What is the function of the tonsils?

   ______________________________________________________________

4. Is there a purpose for the appendix?

   ______________________________________________________________

5. What does the thymus gland do? Where is it located?

   ______________________________________________________________

6. Draw a schematic, sketch, or concept map of the pathway of the lymphatic system as you have seen it in your lab torso models or in the text.

**You may be tested on the names and locations of these anatomical structures.**

# ACTIVITY 2: ENVIRONMENTAL STREAK PLATES AND ANTIBACTERIAL EFFECTIVENESS

Microorganisms, like other life forms, must have adequate nutrients and a favorable environment for growth. In order to culture organisms for testing and study in the lab, we must provide for those needs. A variety of culture media is used and each has a distinctive set of components.

There are three basic forms of culture media: liquid (broth medium), semisolid, and solid. The component which is added to liquid media to create either a semisolid or solid medium is an extract of seaweed called agar. Agar is a complex carbohydrate without nutritional value.

Growth medium can be used in a variety of forms to cultivate microorganisms. We will use agar plates containing a solid nutrient agar medium for our cultures.

Streak plate sampling is done on an agar plate. You will take a sample from the environment and streak it onto the plate. It will then be incubated until next week's lab. You will then be able to see individual colonies of bacteria. We will then take a sample from this and expose it to disinfectants to see which may work against the selected bacteria.

*You can obtain plates from a lab kit, if your school has you use one, or in your lab room.*

## ENVIRONMENTAL STREAK PLATE DIRECTIONS

*Instructions: please read through the instructions **before** you begin!*

1. Obtain 2 agar plates. **Always** leave the lid on unless you are working in the plate.

2. Label the plates with your name, the location from where you obtained your sample, and the date. Be sure you label the ***bottom*** plate around the edges (***not*** through the middle of the plate).

3. Take a cotton swab and dip it into water.

4. Swab an object from the environment (door handle, water fountain, toilet seat, computer, etc.).

5. **Lightly** streak the entire surface of the plate using the swab. The agar is approximately the consistency of gelatin, so be gentle—**do not break the surface of the agar with your swab.**

6. Cover the plate.

7. Place your plates on the metal cart in lab. These will incubate at room temperature until next week.

# ANTIBACTERIAL EFFECTIVENESS DIRECTIONS

## DISINFECTANT EFFECTIVENESS

1. Obtain one of your previously streaked plates (from last week) and identify an area that has a large amount of bacterial growth.

2. Obtain a new agar plate and use a marker to divide the plate into three sections (mark on the **bottom** outer surface of the plate).

3. Label each section with the name of the product to be used in that section.

4. Wet the tip of a cotton swab and gently swab the area of bacterial growth from the previously streaked plate (from last week).

5. Using the swab, lightly streak the new plate covering the entire surface.

6. Sterilize the forceps by passing them through the flame; then, allow them to cool slightly.

7. Use the forceps to get a blank disc, and then dip the disc into the cleaning product of choice.

8. Blot excess product from the blank disc using a paper towel.

9. Drop the disc into the appropriate section of the labeled, newly streaked Petri dish. *Gently* tap the disc into place on the newly streaked agar plate.

10. Repeat Steps 7–9 with the remaining two products you have chosen.

11. Give the treated plates to your instructor or staff member.

12. Hypothesize as to which disinfectant you think will work the best against the environmental bacteria. Write your hypothesis here.

______________________________________________

______________________________________________

______________________________________________

13. Next week check your plates.

    a. How many different bacterial colonies grew on the agar plate?

    ______________________________________________

    b. Of the various environments tested, which two areas grew the most bacteria? Why do you think that may be the case?

    ______________________________________________

    ______________________________________________

    ______________________________________________

**14.** Which disinfectant actually worked the best? ___________________

Was it the one you hypothesized would? ___________________

## AT-HOME ASSIGNMENT

Briefly describe (mode of transmission, organs affected, and symptoms) the following bacterial diseases. You can use cdc.gov to access information on the diseases and organisms that cause them.

## BACTERIAL DISEASES

Whooping cough (*Bordetella pertussis*), Meningococcal meningitis (*Neisseria meningitidis*), Strep throat (*Steptococcus pyogenes*), Pneumococcal pneumonia (*Streptococcus pneumoniae*), Tetanus (*Clostridium tetani*), and Staph skin disease (*Staphylococcus aureus*)

| DISEASE | MODE OF TRANSMISSION | ORGANS AFFECTED | SYMPTOMS | IS THERE A VACCINE? | IS A TREATMENT AVAILABLE? | IS THERE A RESISTANT FORM? |
|---|---|---|---|---|---|---|
|  |  |  |  |  |  |  |
|  |  |  |  |  |  |  |
|  |  |  |  |  |  |  |
|  |  |  |  |  |  |  |
|  |  |  |  |  |  |  |
|  |  |  |  |  |  |  |

## AT-HOME ASSIGNMENT

One major distinction between innate and adaptive immunity is "memory." As you know, the adaptive immune system does have memory, but the innate immune system does not. Please read about a common disease and provide an explanation for what would happen the first and second time someone is infected with that disease under normal conditions. Then explain how this would be different in someone whose adaptive immune system does not function.

# 7

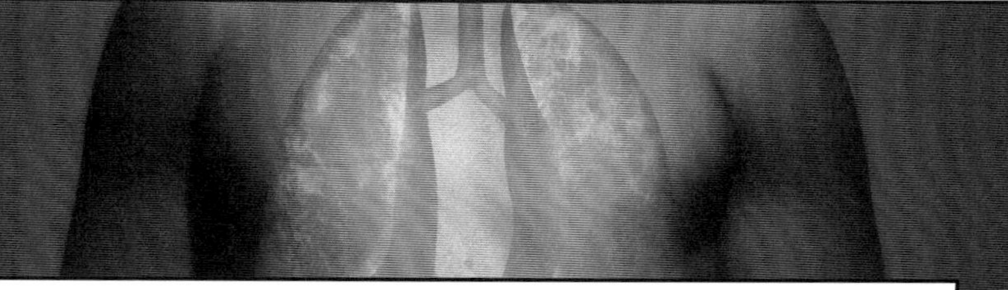

# RESPIRATORY ANATOMY
## PRE-LAB

Name: ________________________    Section: __________    Date: _________

## LEARNING OBJECTIVES

1. Identify the structures and organs of the respiratory system and trace the flow of air through the respiratory system.

2. Identify the organs of the nasal cavity.

3. Identify the areas of the pharynx and surrounding structures.

4. Identify the structures of the larynx.

5. Identify the parts of the trachea and bronchial tree.

6. Identify the components of the respiratory zone.

7. Identify the structures of the lungs, including the pleura that surround the lungs.

8. Identify the major histology on a trachea and lung slide.

*Checklist to complete* **before entering** *the science skills lab (SSL):*

☐ Actively read this packet of information.

☐ Complete the charts, tables or labeling and answer questions using your own words.

☐ Complete the electronic digital pre-lab quiz on Bb by Sunday.

☐ Review the attached anatomy list and take it to lab with you. Jot down key descriptive identifying words that aid in your lab test preparations.

## INTRODUCTION

Respiratory system is divided up into two labs. This lab, which deals predominantly with identifying the anatomy of the major organs and structures of the respiratory system, and the next lab, which will focus on the physiology of the respiratory system.

The word, **respiration,** is derived from the Latin word, respirationem, which means breathing. See the **etymology of respiration** for more information [https://en.wiktionary.org/wiki/respiration]. One of the most obvious features of the respiratory system is the movement of air in and out of the body through ventilation or breathing. Why do we need to breathe? Where does the air go once we breathe it in through our nose or mouth? What causes us to be able to breathe? How does the gas that we breathe in get to our tissues? What might happen to the body if we couldn't breathe? These and other questions will be answered as you go through this lab, the next lab, as well as the readings for the respiratory system. Try not to focus on mere identification during this lab. **Although identifying organs in structures in the lab is an important component of Anatomy and Physiology, try to think about what each part of the respiratory system does, as well as what direction gases would flow in the respiratory system.** One way that you can connect what the different parts of the respiratory system might do is to think about what might happen if that part was missing or if there was a particular disorder. For example, at least one effect of **emphysema** is the destruction over time of the elasticity of the alveoli, which are the exchange surfaces for respiratory gases, like oxygen and carbon dioxide. This would make it harder to get sufficient gases to those exchange surfaces and affect the breathing of the individual. This visual gives you an idea of the function of the alveoli. See if you can find other respiratory disorders that might help you see the importance of different parts of the respiratory system.

## ACTIVITY 1: RESPIRATORY SYSTEM ORGANIZATION

Look at Figure 1 in Unit 3.1, major respiratory structures, to answer the following questions.

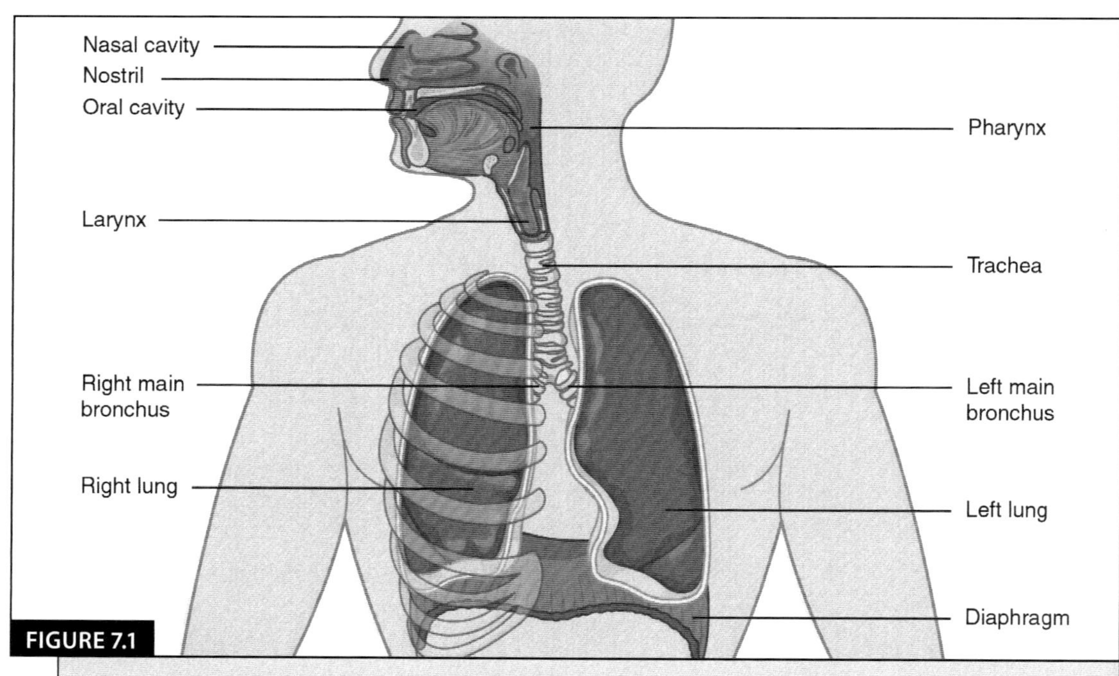


**FIGURE 7.1**

**The major respiratory structures span the nasal cavity to the diaphragm**
Organs and Structures of the Respiratory System [https://cnx.org/contents/FPtK1zmh@8.108:t2sgkCQ-@10/Organs-and-Structures-of-the-Respiratory-System] by OpenStax, (CC BY 4.0)

1.  Name the parts of the upper respiratory system. What is the dividing line between the two areas?

2.  Name the parts of the lower respiratory system.

3.  What parts are considered the conducting zone, and what are some functions of the conducting zone?

4.  What parts are considered the respiratory zone, and what is considered a function of the respiratory zone?

## ACTIVITY 2: STRUCTURE OF THE NASAL CAVITY ——————

Look at Figure 3 in Unit 3.1 What role does the nose play in the conducting zone?

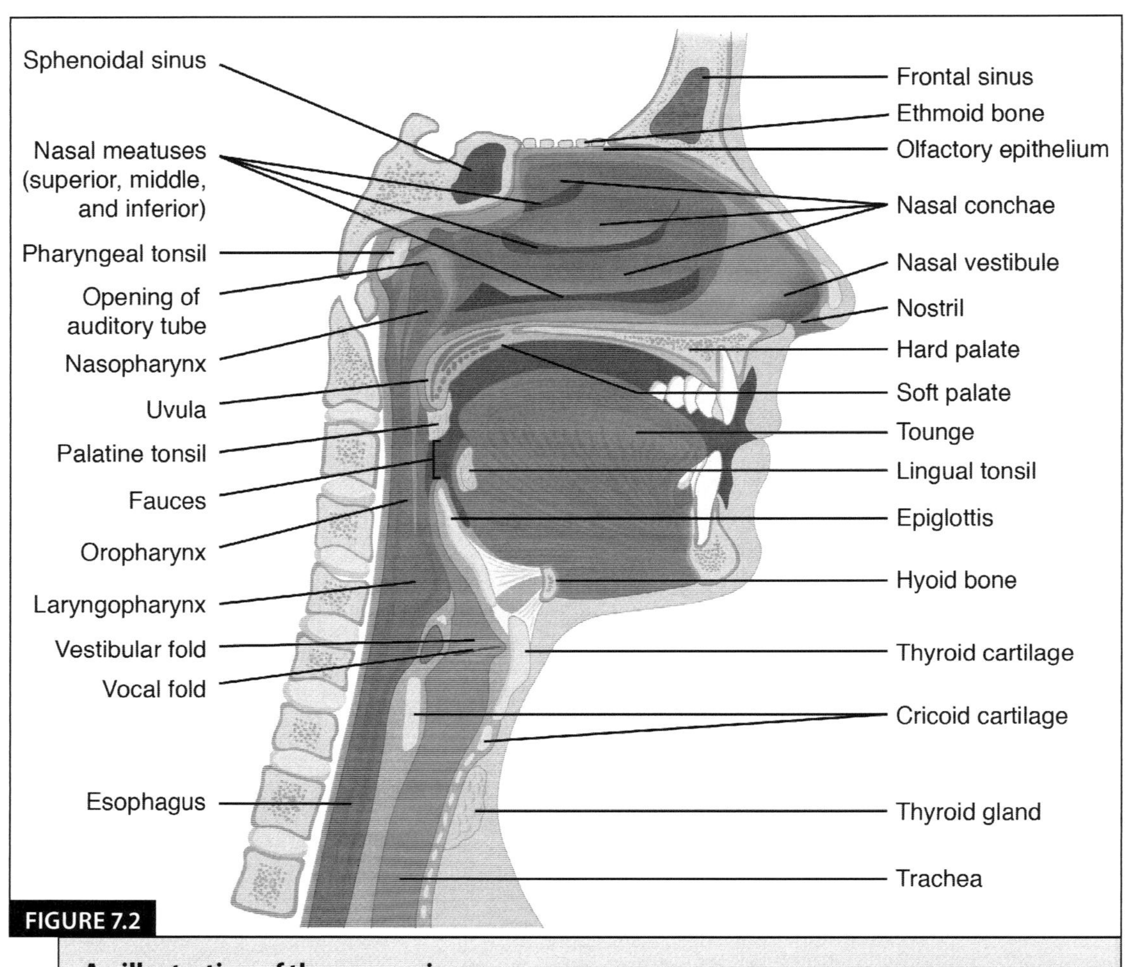


**FIGURE 7.2**

**An illustration of the upper airway**   Organs and Structures of the Respiratory System [https://cnx.org/contents/FPtK1zmh@8.108:t2sgkCQ-@10/Organs-and-Structures-of-the-Respiratory-System] by OpenStax, (CC BY 4.0)

1. Trace the flow of air in the nasal cavity by listing here the structures, in order, that air flows through in the nasal passages. Start with the nostrils or nares.

2. What type of epithelium lines most of the nasal cavity?

3. What is the role of the nasal meatuses?

# ACTIVITY 3: PHARYNGEAL STRUCTURES

Look at Figure 5 in Unit 3.1 What role does the pharynx play in the conducting zone?

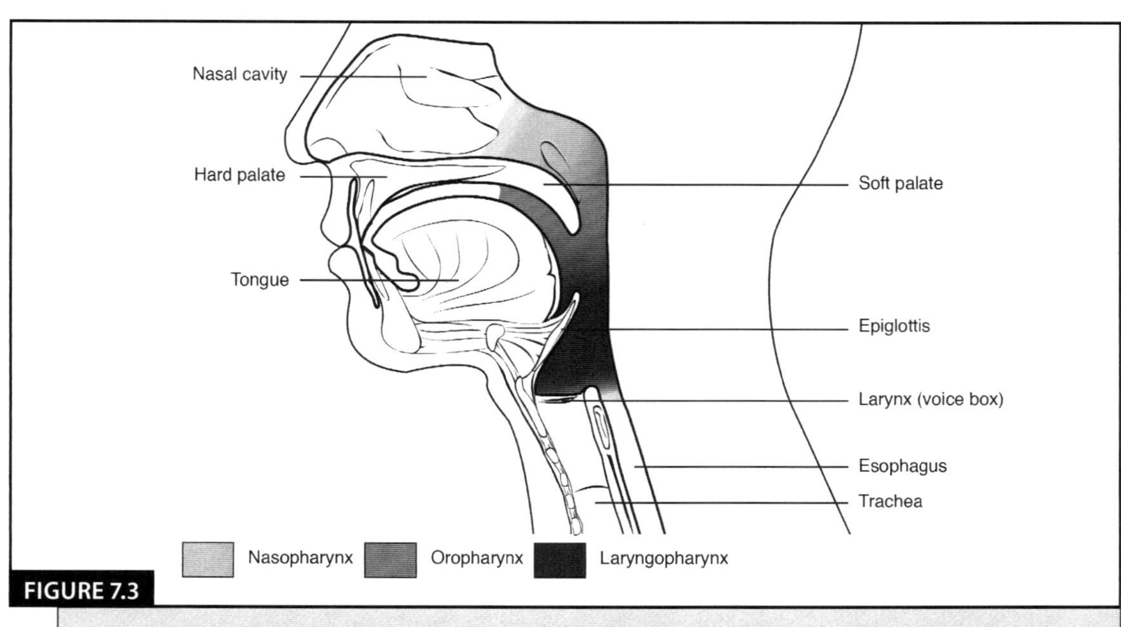


**FIGURE 7.3**

**The pharynx is divided into three regions, seen here in shades of gray.**
Organs and Structures of the Respiratory System [https://cnx.org/contents/FPtK1zmh@8.108:t2sgkCQ-@10/Organs-and-Structures-of-the-Respiratory-System] by OpenStax, (CC BY 4.0)

1.  From your reading and the figure above, name the three parts of the pharynx.

2.  Briefly describe the superior and inferior borders of the three parts of the pharynx.

3.  What three tonsils (one is a tonsil pair) are found around the pharynx? Review: What are purposes of tonsils?

4.  Do some research. What is the vallecula?

# ACTIVITY 4: LARYNGEAL STRUCTURES

Look at Figure 4 and 5 in Unit 3.1. What role does the larynx play in the conducting zone?

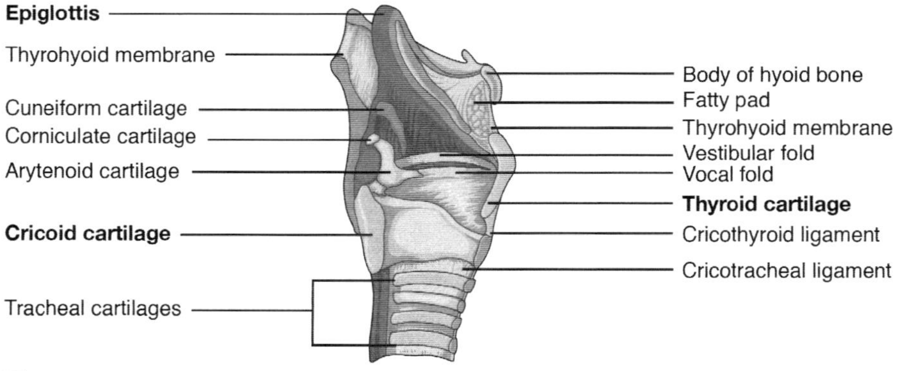

**FIGURE 7.4**

**The larynx extends from the laryngopharynx and the hyoid bone to the trachea**

Organs and Structures of the Respiratory System [https://cnx.org/contents/FPtK1zmh@8.108:t2sgkCQ-@10/Organs-and-Structures-of-the-Respiratory-System] by OpenStax, (CC BY 4.0)

1.  Name the cartilages associated with the larynx. Which is considered the largest?

2.  What is the opening into the trachea (at the voice box or larynx) called? [Hint: The epiglottis folds down over it.]

3.  What is the difference between the vocal folds and the vestibular folds?

# ACTIVITY 5: TRACHEA AND BRONCHIAL TREE

Look at Figure 8 in Unit 3.1. What role does the trachea and bronchial tree play in the conducting zone?

1.  What type of tissue is found in the tracheal cartilages?

    What is the purpose of these cartilages?

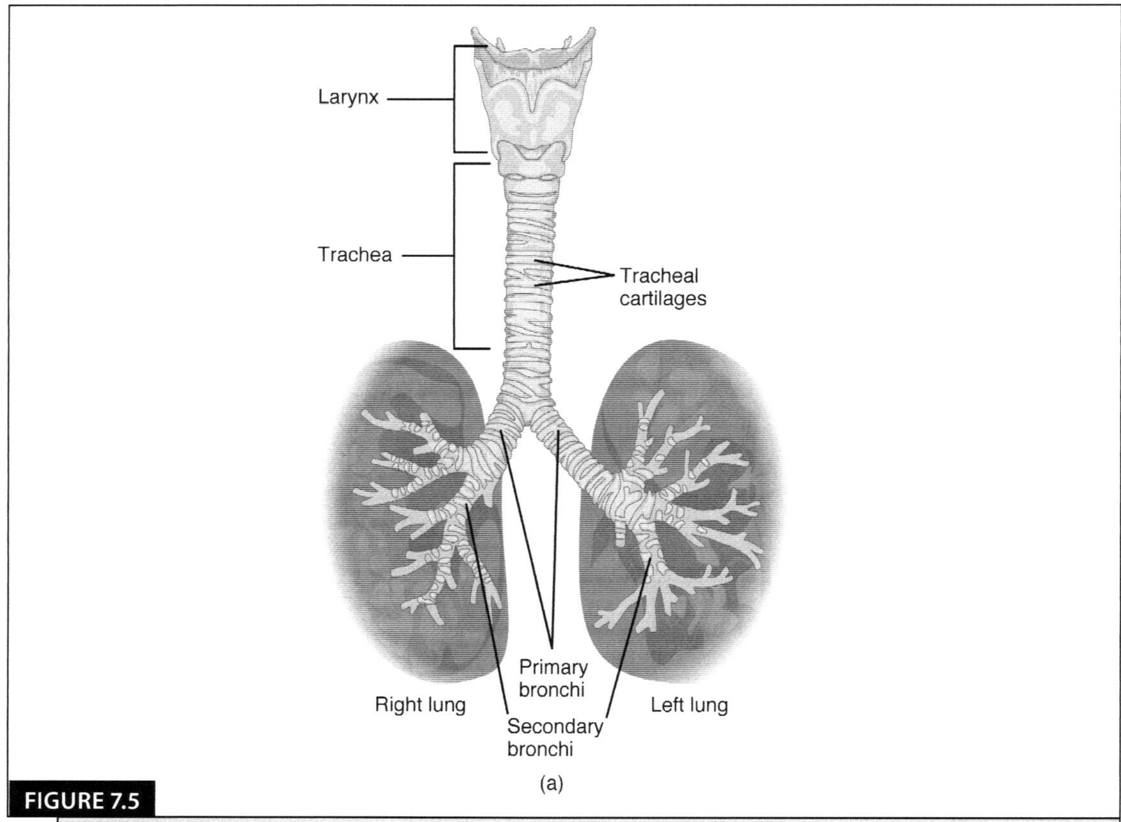


**FIGURE 7.5**

**The tracheal tube is formed by stacked, C-shaped pieces of hyaline cartilage.**
Organs and Structures of the Respiratory System [https://cnx.org/contents/FPtK1zmh@8.108:t2sgkCQ-@10/
Organs-and-Structures-of-the-Respiratory-System] by OpenStax, (CC BY 4.0)

2. What type of tissue lines the inside of the trachea and the primary bronchi?

3. Which side of the respiratory bronchial tree has three secondary or lobar bronchi?

4. What organ is located posterior to the trachea?

## ACTIVITY 6: RESPIRATORY ZONE

Look at Figure 9 and 10 in Unit 3.1. What is the respiratory zone?

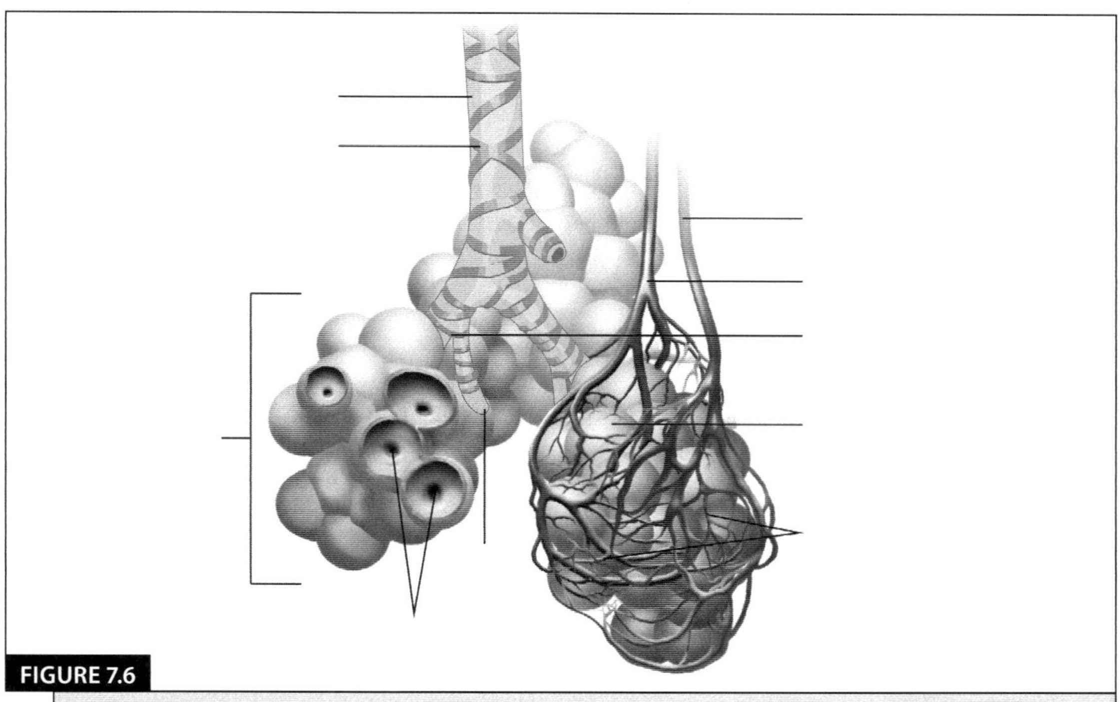

**FIGURE 7.6**

**Alveoli and the respiratory bronchiole**   Organs and Structures of the Respiratory System [https://cnx.org/contents/
FPtK1zmh@8.108:t2sgkCQ-@10/Organs-and-Structures-of-the-Respiratory-System] by OpenStax, (CC BY 4.0)

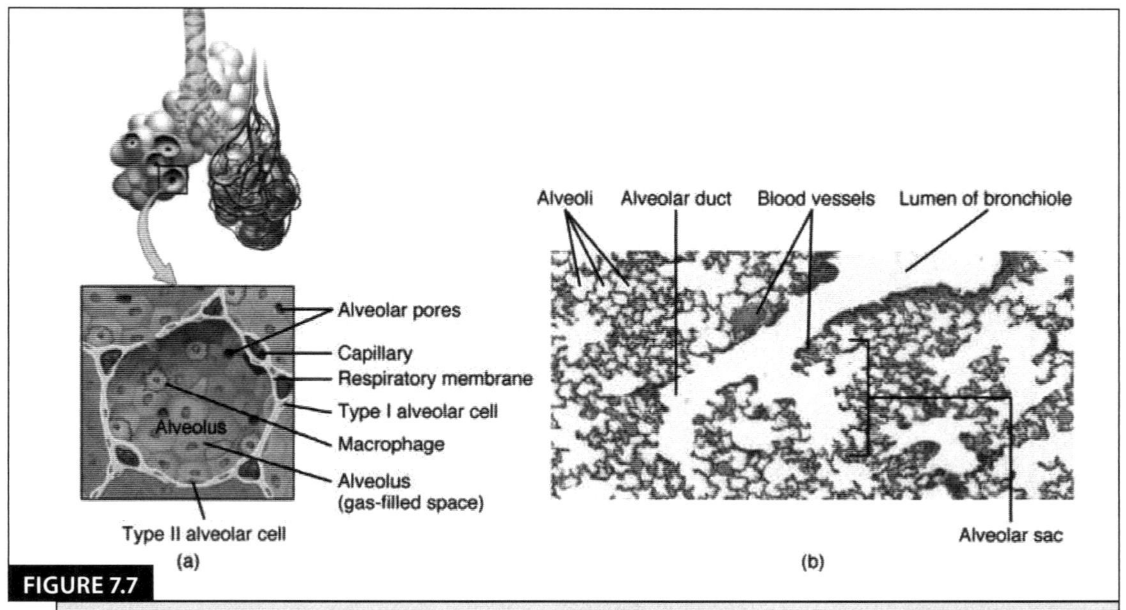


**FIGURE 7.7**

**(a) The alveolus is responsible for gas exchange; (b) A micrograph shows the alveolar
structures within lung tissue, LM × 178.**   Structures of the Respiratory Zone [https://cnx.org/contents/
FPtK1zmh@8.108:t2sgkCQ-@10/Organs-and-Structures-of-the-Respiratory-System] by OpenStax, (CC BY 4.0)
(Micrograph provided by the Regents of University of Michigan Medical School ©2012)

1. What are the components of the respiratory zone? Trace the flow of air, in order, from the tertiary bronchi toward the exchange surfaces in the respiratory zone.

2. Describe in your own words the shape of the alveolar sacs and alveoli. Why are they shaped like this?

3. What type of epithelium is associated with the lining of the alveoli?

4. What cell type secretes surfactant and what is the purpose of **surfactant?**

# ACTIVITY 7: LUNGS AND PLEURA

*Read about the role of the lungs and the pleura in Unit 3.1.*

A major organ of the respiratory system, each lung houses structures of both the conducting and respiratory zones. The main function of the lungs is to perform the exchange of oxygen and carbon dioxide with air from the atmosphere. To this end, the lungs exchange respiratory gases across a very large epithelial surface area—about 70 square meters—that is highly permeable to gases.

The pleurae perform two major functions. They produce pleural fluid and create cavities that separate the major organs. Pleural fluid is secreted by mesothelial cells from both pleural layers and acts to lubricate their surfaces. This lubrication reduces friction between the two layers to prevent trauma during breathing and creates surface tension that helps maintain the position of the lungs against the thoracic wall. This adhesive characteristic of the pleural fluid causes the lungs to enlarge when the thoracic wall expands during ventilation, allowing the lungs to fill with air. The pleurae also create a division between major organs that prevents interference due to the movement of the organs, while preventing the spread of infection.

1. Which lung has two lobes?

2. The visceral pleura is attached to what structure?

3. The parietal pleura is attached to what structure?

4. What functional part of the nervous system regulates the diameter of the airways?

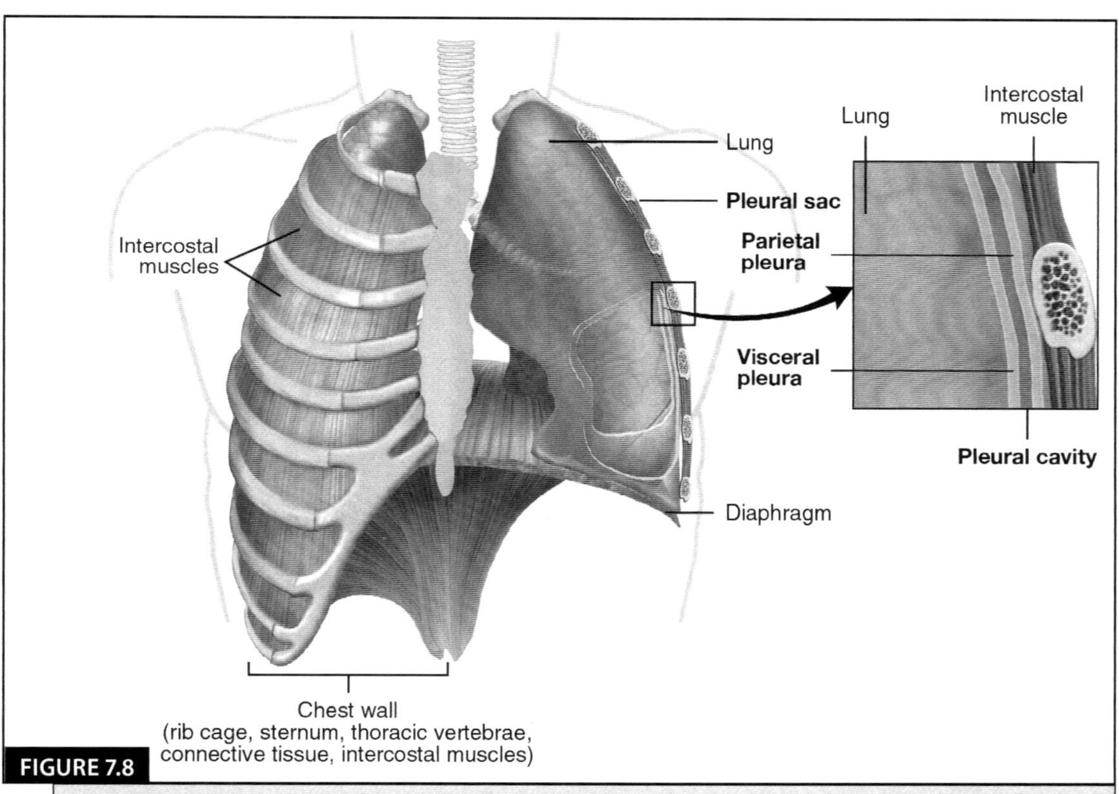


**FIGURE 7.8**

**Parietal and visceral pleurae of the lungs**

The Lungs [https://cnx.org/contents/FPtK1zmh@8.108:udJfuR_E@5/The-Lungs] by OpenStax, (CC BY 4.0)

# 7

# RESPIRATORY ANATOMY
## IN-LAB ACTIVITIES

Name: _________________________    Section: _________    Date: _________

*The activities in this lab are associated with the lecture materials in Unit 3.1 on the respiratory system.*

## LEARNING OBJECTIVES

1. Identify the structures and organs of the respiratory system.
2. Outline the functions of respiratory organs.
3. Follow the route of air into and out of the lungs.

## INSTRUCTIONS FOR STUDENTS

Read the lecture material and complete all pre-lab activities before coming to lab. Read and review the lab activities beforehand. When you come to lab, follow this handout and make sure the lab instructor checks off your work once you finish.

## INTRODUCTION AND BACKGROUND: HUMAN RESPIRATORY SYSTEM AND RESPIRATORY ANATOMY

The major organs of the respiratory system function primarily to provide oxygen to body tissues for cellular respiration, remove the waste product carbon dioxide, and help to maintain acid-base balance. Portions of the respiratory system are also used for non-vital functions, such as sensing odors and speaking.

Structurally, the respiratory system can be divided into an **upper respiratory system** and a **lower respiratory system.** The upper respiratory system consists of the **nose** and **nasal cavity** as well as the **pharynx.** The lower respiratory system consists of the **larynx, trachea,** the **bronchial tree,** and the **lungs.**

Functionally, the respiratory system can be divided into a **conducting zone** and a **respiratory zone.** The conducting zone of the respiratory system includes the organs and structures not directly involved in gas exchange and includes the nose, pharynx, larynx, trachea, bronchi, and bronchioles to terminal bronchioles. While the conducting zone does not participate in gas exchange between the lungs and the blood, it does have several important functions in terms of the quality of the air that reaches the respiratory zone. Some of these functions include filtering, warming, and humidifying the air to more closely match the conditions of the exchanges surfaces and minimize damage to them. Gas exchange occurs in the respiratory zone which includes respiratory bronchioles, alveolar ducts, alveolar sacs, and alveoli.

## ACTIVITY 1: RESPIRATORY SYSTEM ORGANIZATION

On models in the lab, identify the following: **nasal cavity, nostrils, oral cavity, pharynx, larynx, trachea, right/left primary bronchi, right/left lung, and diaphragm.**

Label these in the torso below, drawing in any needed organs.

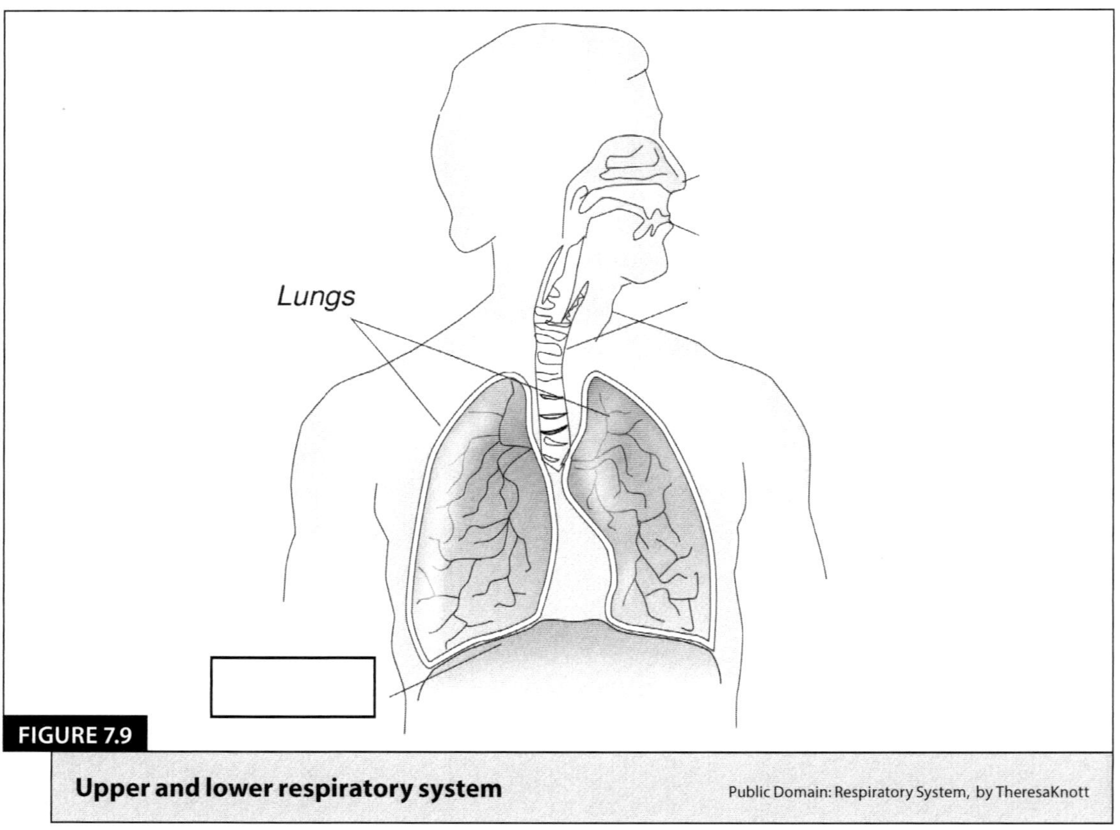


**FIGURE 7.9**

**Upper and lower respiratory system**

Public Domain: Respiratory System, by TheresaKnott

# Upper Respiratory System

## ACTIVITY 2: STRUCTURE OF THE NASAL CAVITY

Air enters into the respiratory system from the external environment through the nostrils, also called the **external nares,** which open into the **nasal cavity.** The nasal cavity is separated into a left and right section by the **nasal septum,** which is formed anteriorly by parts of the septal cartilage and posteriorly by the bony portion that consists of the perpendicular plate of the ethmoid bone and the vomer bones. Each lateral wall of the nasal cavity has three bony projections called the **superior, middle, and inferior nasal conchae.** The inferior conchae are separate bones, whereas the superior and middle conchae are portions of the ethmoid bone. Conchae serve to increase the surface area of the nasal cavity and to disrupt the flow of air as it enters the nose, causing air to bounce along the epithelium where it is cleaned and warmed. The conchae are separated from each other and the **hard palate** by the **superior, middle, and inferior meatuses.** The conchae and meatuses also conserve water and prevent dehydration of the nasal epithelium by trapping water during exhalation. The floor of the nasal cavity is composed of the palate. The **hard palate** at the anterior region of the nasal cavity is composed of bone. The **soft palate** at the posterior portion of the nasal cavity consists of muscle tissue and a flap called the **uvula,** which prevents food and liquid from getting into the airways. Air exits the nasal cavities via the **internal nares** and moves into the pharynx. Below the hard and soft palate is the **oral cavity.** Inside the oral cavity is a muscular organ called the tongue that is involved in propelling food into the pharynx during swallowing. This tongue is attached to the floor of the oral cavity by a structure called the **lingual frenulum.** Anterior to the tongue are the teeth that are involved in chewing and processing food that enters the oral cavity. It is attached to the lips through what are called the **labial frenulum** and are protected from disease by the **gingiva** or gums.

Label the following illustration and *identify on the models in the lab* the structure and function of the following parts: **Nasal septum, external nares, nasal conchae (superior, middle, and inferior), nasal meatuses (superior, middle, and inferior), internal nares, hard palate, soft palate, uvula, tongue, lingual frenulum, labial frenulum, and gingiva**

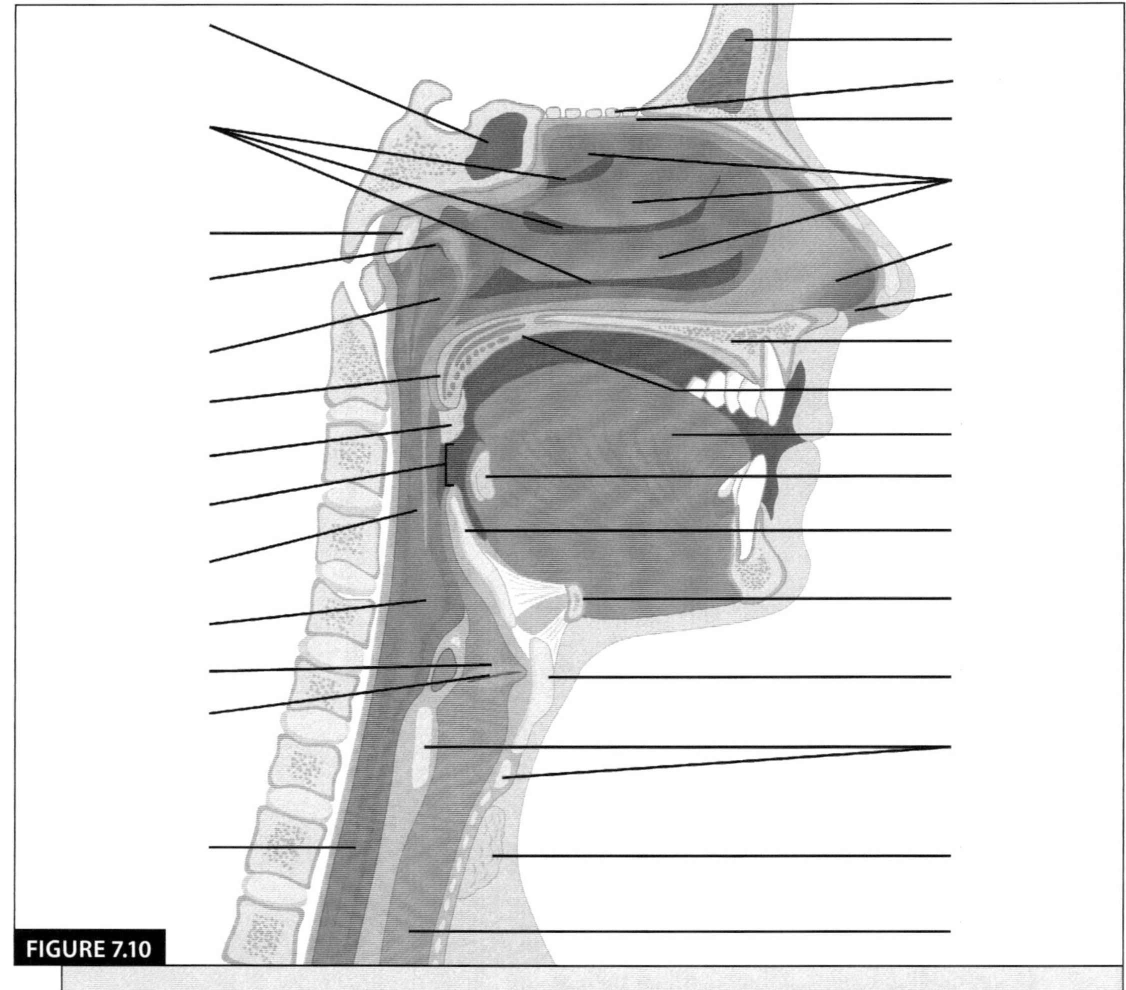

**FIGURE 7.10**

**Upper respiratory system seen in midsagittal section.** Labeled by students.
Organs and Structures of the Respiratory System [https://cnx.org/contents/FPtK1zmh@8.108:t2sgkCQ-@10/
Organs-and-Structures-of-the-Respiratory-System] by OpenStax, (CC BY 4.0)

# ACTIVITY 3: PHARYNGEAL STRUCTURES

The pharynx is a tube that is shared by both the respiratory and digestive system and leads either into the larynx or the esophagus. The pharynx is divided into three major regions: the **nasopharynx,** the **oropharynx,** and the **laryngopharynx.** The nasopharynx comes off of the internal nares and contains the pharyngeal tonsils, which serve to filter incoming air from possible pathogens. Both the **uvula** and **soft palate** move during swallowing, swinging upward to close off the nasopharynx to prevent ingested materials from entering the nasal cavity. In addition, auditory (Eustachian) tubes that connect to each middle ear cavity open into the nasopharynx. The oropharynx is a passageway for both air and food. The oropharynx is bordered superiorly by the nasopharynx and anteriorly by the oral cavity. The oropharynx contains two distinct sets of tonsils, the **palatine** and **lingual tonsils.** The laryngopharynx is inferior to the oropharynx and posterior to the larynx. It continues the route for ingested material and air until its inferior end, where the digestive and respiratory systems diverge. Another feature that can be seen in the general vicinity of the pharynx is the **vallecula.** This space is located between the **epiglottis** (discussed with the larynx) and the base of the tongue and is a common landmark for intubation (inserting a tube into the airways).

Identify on the following illustration **the nasopharynx, oropharynx, vallecula, epiglottis, soft palate, pharyngeal tonsils, entrance to the auditory (Eustachian) tube, palatine tonsils, lingual tonsils,** and **laryngopharnx.** Use sagittal heads in lab to identify these anatomical areas for the lab test.

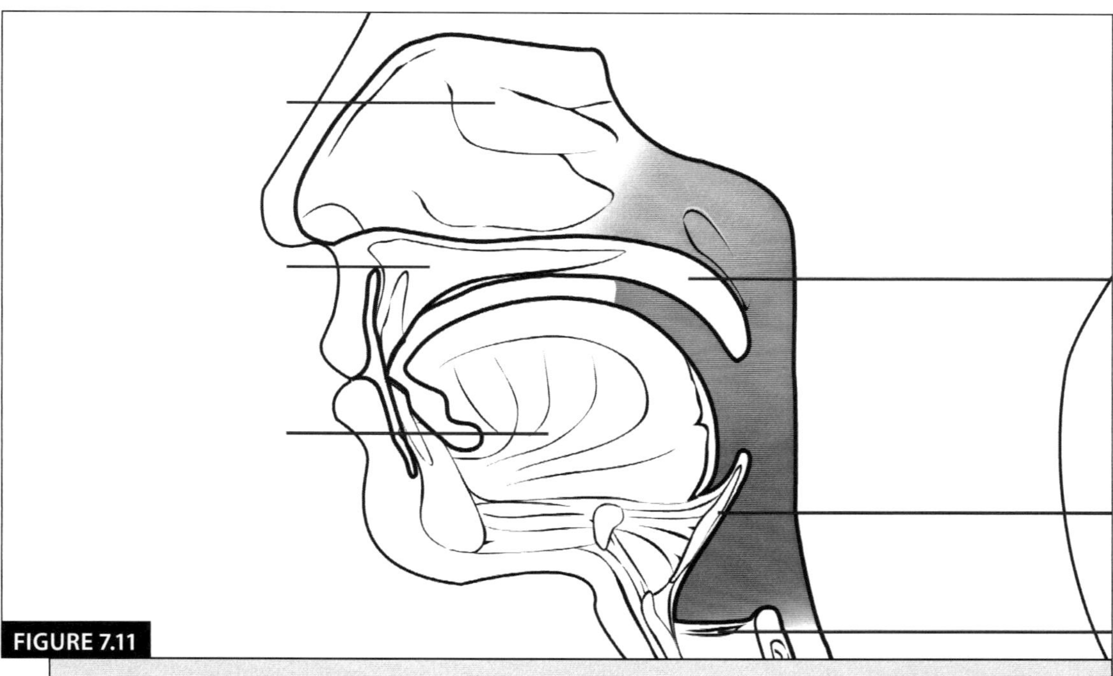

**FIGURE 7.11**

**Upper respiratory system.** Shaded areas mark the nasopharynx, the oropharynx, and the laryngopharynx to be labeled, along with the tonsils, the epiglottis, and the opening to the auditory tube. Organs and Structures of the Respiratory System [https://cnx.org/contents/FPtK1zmh@8.108:t2sgkCQ-@10/Organs-and-Structures-of-the-Respiratory-System] by OpenStax, (CC BY 4.0)

# Lower Respiratory System

## ACTIVITY 4: LARYNGEAL STRUCTURES

The larynx is a cartilaginous structure inferior to the laryngopharynx that connects the pharynx to the trachea. The structure of the larynx is formed by several pieces of cartilage. Three large cartilage pieces—the thyroid cartilage, **epiglottis,** and cricoid cartilage—form the major structure of the larynx. The thyroid cartilage is the largest piece of cartilage that makes up the larynx. The thyroid cartilage consists of the laryngeal prominence, or "Adam's apple," which tends to be more prominent in males. It is connected to the next largest cartilage, the **cricoid cartilage,** by the **cricothyroid ligament.** Three smaller, paired cartilages—the arytenoids, corniculates, and cuneiforms—attach to the epiglottis and the vocal cords. When in the "closed" position, the unattached end of the epiglottis rests on the glottis. The glottis is composed of the vestibular folds, the vocal folds, and the space between these folds. The **vocal folds** vibrate to produce sound and are sometimes called the true vocal cords. The **vestibular folds** are inelastic and are sometimes referred to as the false vocal cords.

On the following illustrations and *models in the lab* identify the following parts: **thyroid cartilage, cricoid cartilage, arytenoid cartilage, corniculate cartilage, cricothyroid ligament, vestibular folds, vocal folds (true vocal cords), glottis, epiglottis,** and **laryngeal prominence**

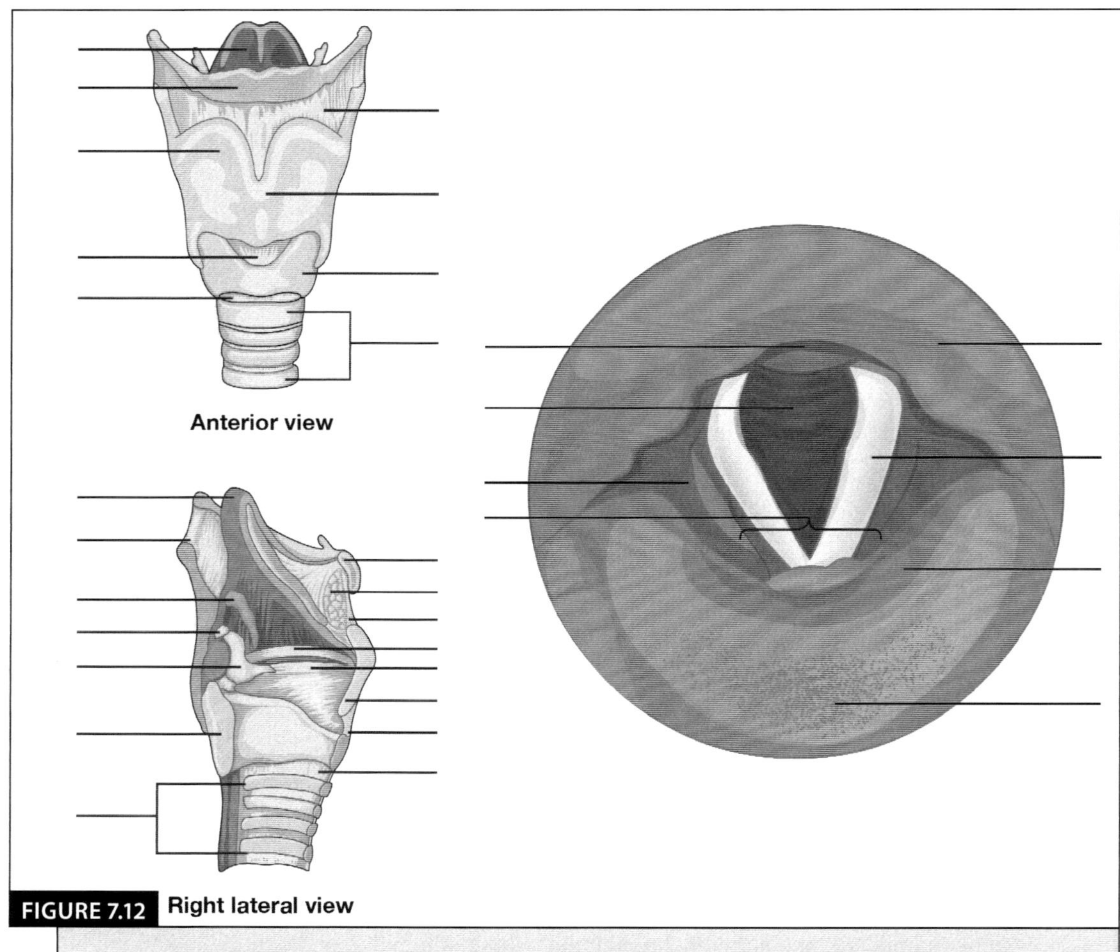


**FIGURE 7.12** Right lateral view

**Anterior and lateral view of the Larynx, unlabeled.** Superior view of the internal larynx showing the epiglottis and the vocal folds. Organs and Structures of the Respiratory System [https://cnx.org/contents/FPtK1zmh@8.108:t2sgkCQ-@10/Organs-and-Structures-of-the-Respiratory-System] by OpenStax, (CC BY 4.0)

# ACTIVITY 5: STRUCTURES OF THE TRACHEA AND BRONCHIAL TREE

The trachea (windpipe) extends from the larynx toward the lungs. The trachea is formed by several C-shaped pieces of hyaline cartilage (**tracheal cartilages**) that are connected by dense connective tissue. The rings of cartilage provide structural support and prevent the trachea from collapsing. The trachea is lined with pseudostratified ciliated columnar epithelium, which is continuous with the larynx. The esophagus borders the trachea posteriorly. The trachea branches into the right and left **primary** bronchi at the **carina.** Rings of cartilage, similar to those of the trachea, support the structure of the bronchi and prevent their collapse. The primary bronchi enter the lungs at the **hilum,** a concave region where blood vessels, lymphatic vessels, and nerves also enter the lungs. The bronchi continue to branch into a structure that looks like a tree and is referred to as a **bronchial tree.** Branches off of this bronchial tree include the **secondary (lobar) bronchi,** which go into each lobe of the lungs, and the **tertiary (segmental) bronchi,** which lead to different lung segments within different lobes of the lungs. A bronchiole branches from the tertiary bronchi. Bronchioles further branch until they become the tiny terminal bronchioles, which lead to the structures of gas exchange.

Identify the following parts on the following illustration as well on models in the lab: **tracheal cartilages, carina, right mainstem (primary) bronchus, left mainstem (primary) bronchus, lobar (secondary) bronchi, segmental (tertiary) bronchi,** and **hilum (hilus).** (Use the bronchi tree and the lungs with diaphragm models.)

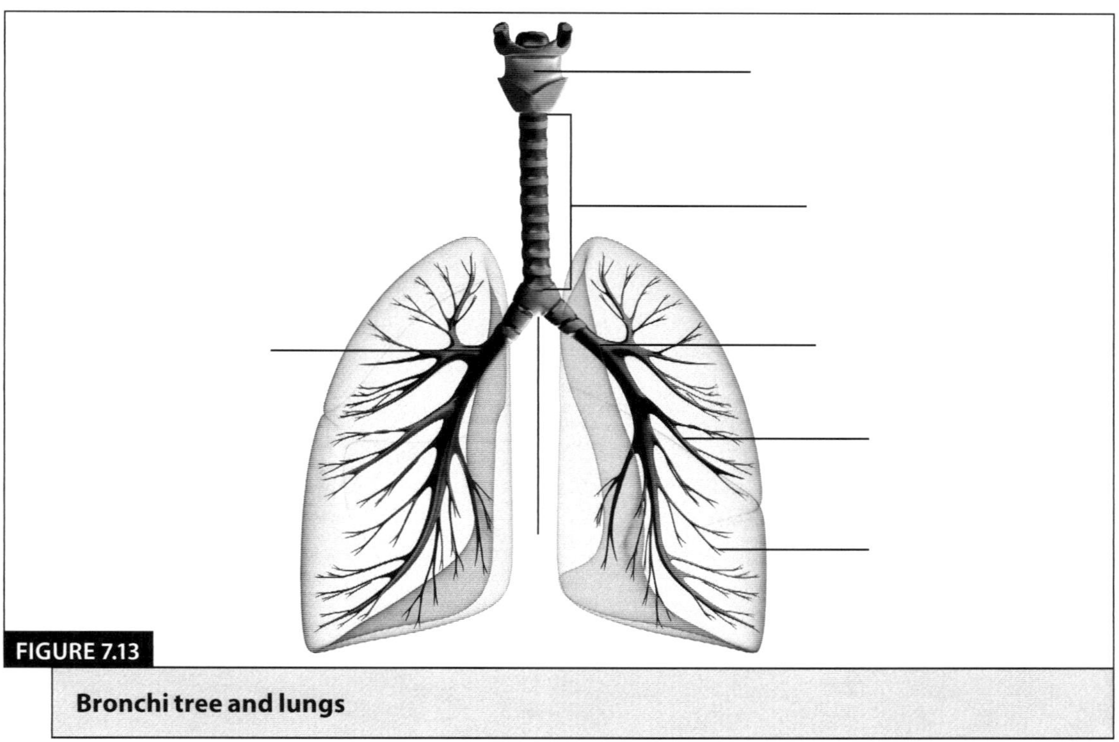

**FIGURE 7.13**

**Bronchi tree and lungs**

## ACTIVITY 6: RESPIRATORY ZONE

In contrast to the conducting zone, the respiratory zone includes structures that are directly involved in gas exchange. The respiratory zone begins where the terminal bronchioles join a respiratory bronchiole , the smallest type of bronchiole, which then leads to an alveolar duct, opening into a cluster of **alveoli.**

Identify the following on the illustration: **respiratory bronchiole, alveolar sacs, alveoli, and alveolar capillaries.**

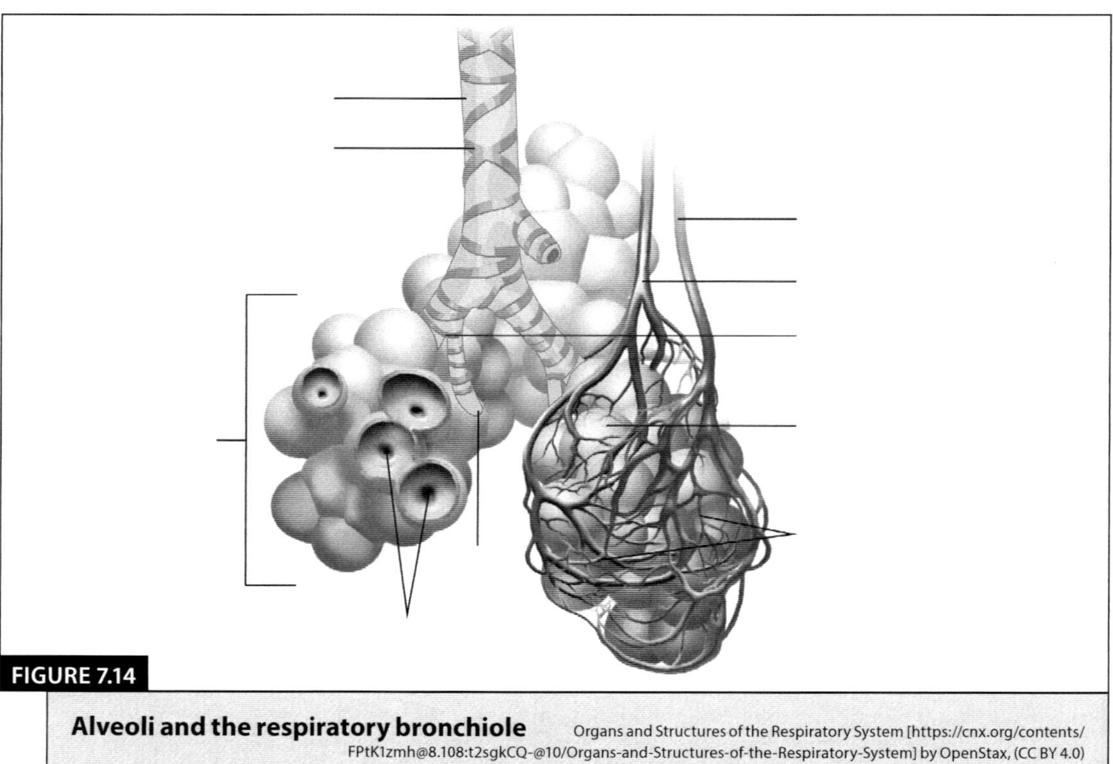

**FIGURE 7.14**

**Alveoli and the respiratory bronchiole** Organs and Structures of the Respiratory System [https://cnx.org/contents/ FPtK1zmh@8.108:t2sgkCQ-@10/Organs-and-Structures-of-the-Respiratory-System] by OpenStax, (CC BY 4.0)

## ACTIVITY 7: LUNGS AND PLEURA

A major organ of the respiratory system, each lung houses structures of both the conducting and respiratory zones. The lungs are enclosed by the **pleurae,** which are attached to the mediastinum. The right lung is shorter and wider than the left lung, and the left lung occupies a smaller volume than the right. The cardiac notch is an indentation on the surface of the left lung, and it allows space for the heart. The **apex** of the lung is the superior region, whereas the **base** is the opposite region near the **diaphragm.** Each lung is composed of smaller units called **lobes.** Fissures separate these lobes from each other. The right lung consists of three lobes: **the superior, middle, and inferior lobes.** The superior and middle lobe are separated from each other by a fissure called the **horizontal fissure,** while the inferior lobe is separated from the other two by the **oblique**

**fissure.** The left lung consists of two lobes: the **superior and inferior lobes.** These two lobes are separated from each other by the oblique fissure. Although the left lung only has two lobes, there is a section of the superior lobe of the left lung that corresponds with the middle lobe of the right lung and carries many structural similarities to the middle lobe called the **lingular segment.** The major function of the lungs is to perform gas exchange, which requires blood from the pulmonary circulation. The pulmonary artery is an artery that arises from the pulmonary trunk and carries deoxygenated, arterial blood to the alveoli. As they near the alveoli, the pulmonary arteries become the pulmonary capillary network. Once the blood is oxygenated, it drains from the alveoli by way of multiple **pulmonary veins,** which exit the lungs through the hilum.

On the following illustration as well as the models in the lab identify the following: **apex, base, hilus, right lung (superior, middle, and inferior lobe), hilus, bronchi (primary, secondary, and tertiary), left lung, superior (lingual) segment, inferior lobe, horizontal fissure, oblique fissures,** and **cardiac notch.** (Use the lungs with diaphragm model in the lab.)

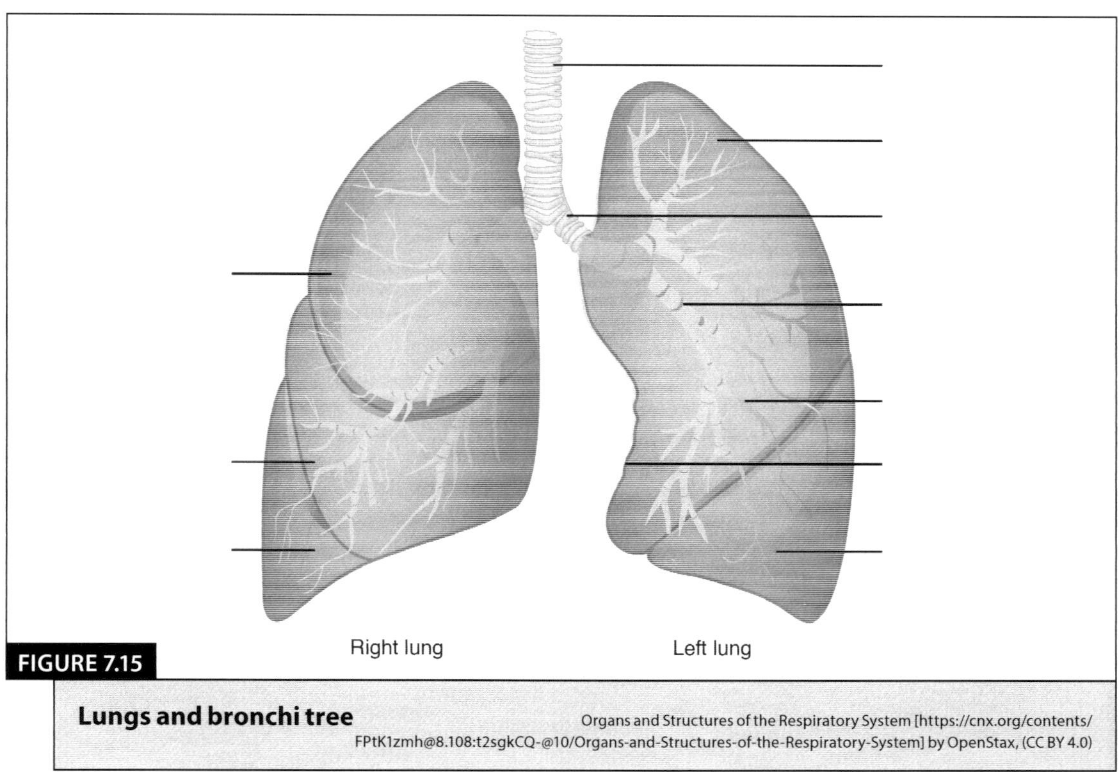


**FIGURE 7.15**

**Lungs and bronchi tree**

Organs and Structures of the Respiratory System [https://cnx.org/contents/FPtK1zmh@8.108:t2sgkCQ-@10/Organs-and-Structures-of-the-Respiratory-System] by OpenStax, (CC BY 4.0)

Each lung is enclosed within a cavity that is surrounded by a serous membrane called the pleura. The pleurae consist of two layers. The **visceral pleura** is the layer that is superficial to the lungs and extends into and lines the lung fissures. In contrast, the **parietal pleura** is the outer layer that connects to the thoracic wall, the mediastinum, and the diaphragm. The **pleural cavity** is the space between the visceral and parietal layers. Within the pleural cavity is a watery fluid called the pleural fluid. Because of the attraction of the water molecules within the pleural fluid, the visceral pleura and parietal pleura are stuck to each other much like a coverslip to a slide. This property will become important to note when understanding how breathing occurs in the next lab. Please review the video about serous membranes in Unit 3.1 What is the pleura of the lungs?

For the following illustration identify the following: **visceral pleura, pleural cavity with pleural fluid, and parietal pleura**

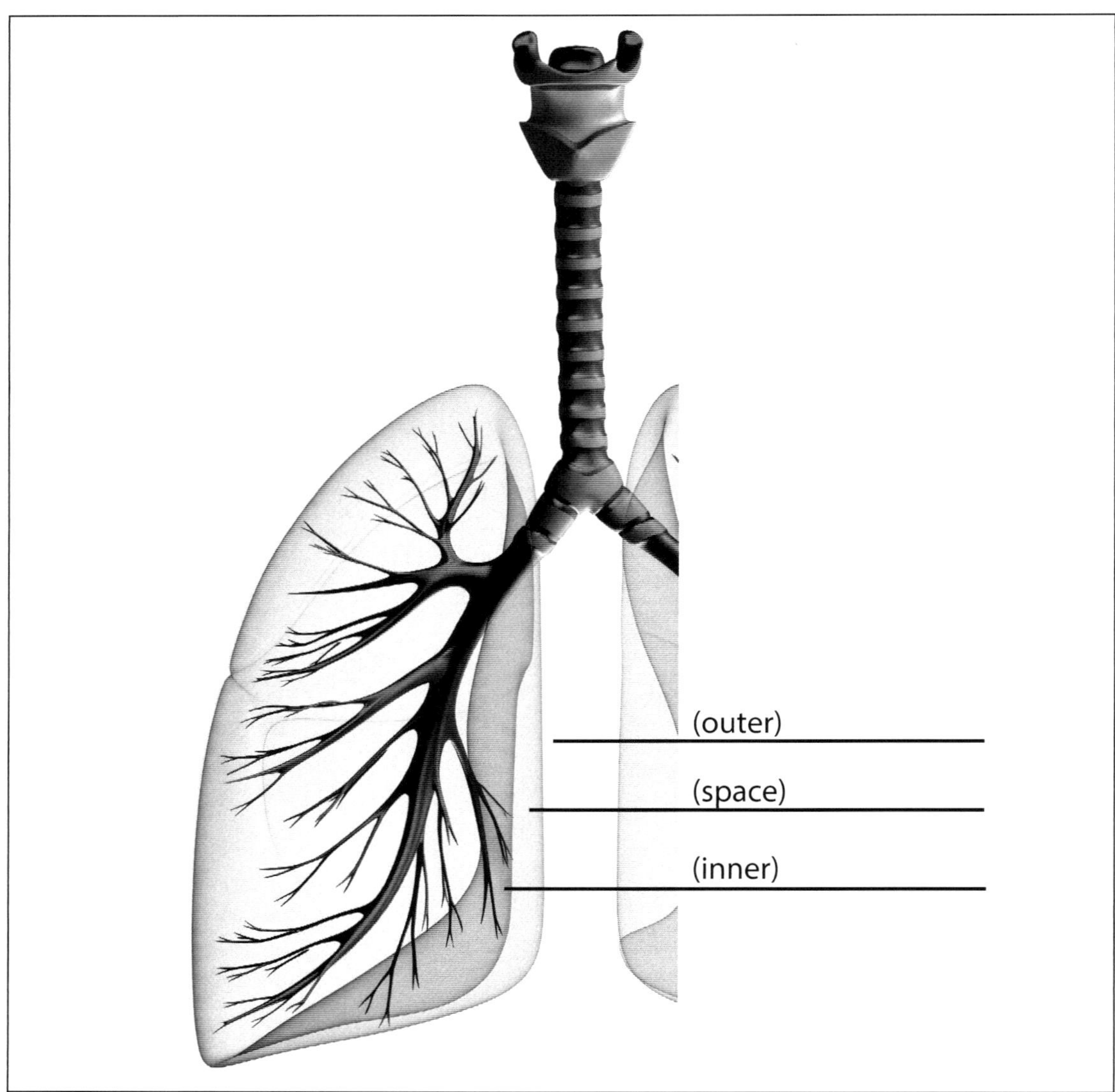

## ACTIVITY 8: HISTOLOGY OF THE RESPIRATORY SYSTEM

*Review the trachea in Unit 3.1 and identify the following on a slide of the* **trachea:**

**Tracheal cartilage (hyaline cartilage**—looks like eyeballs or cat eyes) **and ciliated pseudostratified columnar epithelium,** which surrounds the **lumen**

- Draw what you see in the space that is provided.

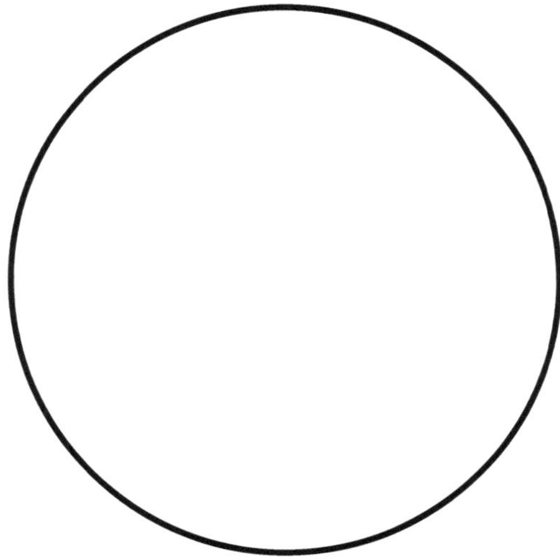

*Review the respiratory zone in Unit 3.1 and identify the following on a slide of the* **lung:**

- **bronchiole, alveoli, and alveolar capillary membrane.**
- Draw what you see in the space provided.

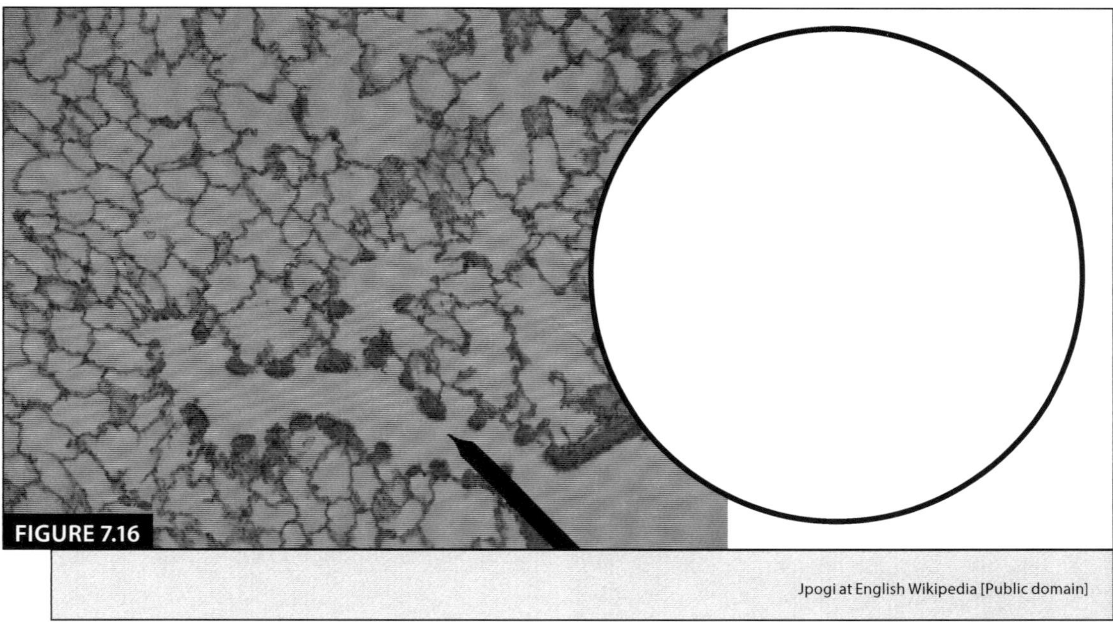



Jpogi at English Wikipedia [Public domain]

## EXTENSION QUESTIONS

1. Name the components that make up the conducting zone and the respiratory zone. What are the primary functions of the conducting zone and respiratory zone?

2. Trace the flow of air from the outside environment to the gas exchange surfaces in the alveoli. Include in your answer which components are part of the conducting zone and which components are part of the respiratory zone.

3. Cystic fibrosis is an inherited disorder caused by a faulty chloride channel. This faulty chloride channel leads to the production of thick mucus rather than thin mucus. Explain what effect cystic fibrosis would have on the function of the respiratory system.

4. What vital role do the pleurae have in respiratory function?

   What is the role of the pleural fluid?

**5.** What is the purpose of the vallecula? Describe where it is located.

**6.** What specific type of tissue is associated with the following locations?

    **a.** lining of the nasal cavity

    **b.** in the oropharynx

    **c.** making up the tracheal cartilages

    **d.** lining the alveoli

    **e.** lining the trachea

**7. True or False.** All speech is produced from the vibrations of the vocal folds. **Explain.**

**8.** In which primary bronchus, the right or left, are objects more likely to get into? What is the reason for this?

9. What typically occurs when someone is choking?

10. Describe how the following disorders affect the structure and function of the respiratory system.

   **a.** asthma

   **b.** COPD

   **c.** smoking

# 8

# RESPIRATORY PHYSIOLOGY
## PRE-LAB

Name: _________________________  Section: ___________  Date: _________

## LEARNING OBJECTIVES

1. Identify the respiratory muscles and describe their function.

2. Demonstrate the pressure and volume changes that occur during breathing.

3. Recognize the major lung volumes and capacities and how they change before and after exercise.

4. Have knowledge of the equation for carbon dioxide conversion into bicarbonate and understand what effect this would have on pH before and after exercise.

5. Recognize the oxygen-hemoglobin saturation curve and how this is affected by the partial pressure of oxygen and pH.

*Checklist to complete* **before entering** *the science skills lab (SSL):*

☐ Actively read this packet of information.

☐ Complete the charts, tables or labeling and answer questions using your own words.

☐ Complete the electronic digital pre-lab quiz on Bb by Sunday.

☐ Review the attached anatomy list and take it to lab with you. Jot down key descriptive identifying words that aid in your lab test preparations.

## INTRODUCTION

The respiratory system and along with the cardiovascular system are the two major organ systems of the body that are responsible for the exchange of gases such as oxygen and carbon dioxide. Both oxygen and carbon dioxide have important roles to play in the functioning of the human body. Oxygen is used by the cells during a series of reactions called **cellular respiration** *that allows cells to convert the energy from organic molecules, such as glucose,* into a useable form of energy stored predominantly in the adenosine triphosphate (ATP) molecule. These important reactions are critical to the survival of the body and take place inside the mitochondria of cells. Without sufficient intake and transport of oxygen to the body cells, these important energy-releasing reactions cannot take place. Carbon dioxide is a waste product of cellular respiration but also has a major role to play in the regulation of acid and bases in the human body. Too much or too little acid in the body has a negative impact on the functioning of the human body.

It is important to understand how oxygen and carbon dioxide are exchanged between the atmosphere and the body; it is also essential to understand how oxygen and carbon dioxide are transported in the body. In the case of carbon dioxide, this is predominantly by converting it to bicarbonate and binding oxygen to hemoglobin. You may have already read about these transport mechanisms when studying the blood in an earlier lab, but this will be expanded more in this lab. Without these transport mechanisms, oxygen and carbon dioxide transport will be compromised.

Respiratory health and functioning is often measured using what are called **lung volumes and lung capacities.** These volumes and capacities are ways of describing how much air can be drawn in and out of the lungs. These volume and capacities change based on activity level, pH, carbon dioxide, oxygen levels, or even by certain respiratory disorders such as emphysema.

In lab, you will be asked to identify the individual respiratory muscles and their specific roles in breathing. You will learn from both the readings, as well as the lab, that when these respiratory muscles contract, they will either increase or decrease the volume inside of the lungs, which will result in changes in pressure inside of the lungs.

The volume and pressure changes are described by a respiratory law called **Boyle's Law.** By understanding this law, you will be able to identify how the lungs are able to bring air in and out of the body.

**Watch the following video for further explanation of Boyle's Law.**

- https://www.youtube.com/watch?v=fH1H4iglW2A

## ACTIVITY 1:

1. **Review** the *major respiratory muscles* found in the abdominal wall and the thorax of the human body by reading the following section of the Open Stax textbook.

   - https://cnx.org/contents/FPtK1zmh@8.108:b3YG6PIp@6/Axial-Muscles-of-the-Abdomina

2. **Read** about the axial muscles of the head and neck, then answer the following questions.

3. **Describe** the area in which you would find each respiratory muscle and identify if the following muscles are involved in inspiration or expiration.

    **a.** Sternocleidomastoid

    **b.** Diaphragm

    **c.** External intercostals

    **d.** Internal intercostals

    **e.** External/internal obliques

    **f.** Rectus abdominis

    **g.** Transverse abdominis

4. **Review** the *major skeletal structures* that make up the thoracic cage and protect the respiratory organs by reading the following section of the **Open Stax textbook.**

   - https://cnx.org/contents/FPtK1zmh@8.108:inB64l2K@4/The-Thoracic-Cage

5. **Identify** and **study** the skeletal and muscular structures in preparation for lab. You will be using models in the lab to practice identifying these structures with a partner.

## ACTIVITY 2:

Watch the video on **Boyle's Law** and answer the following questions:

- https://www.youtube.com/watch?v=fH1H4iglW2A

1. What happens to pressure as volume changes?

2. What respiratory muscles are associated with increasing volume? Does this cause inspiration or expiration?

3. What respiratory muscles are associated with decreasing volume? Does this cause inspiration or expiration?

## ACTIVITY 3:

**Describe** how the following lung volumes and capacities are measured or calculated using a *spirometer.*

1. Tidal volume

2. Expiratory reserve volume

3. Vital capacity

4. Inspiratory reserve volume

# ACTIVITY 4:

1. Write the chemical equation that involves the reaction of carbon dioxide with water. What enzyme catalyzes this reaction?

2. If carbon dioxide levels increase, what might happen to pH?

# ACTIVITY 5:

***Read Unit 3.3 in Odigia text.*** *Answer the following questions.*

1. What is meant by the percent oxygen saturation of hemoglobin?

2. Define partial pressure.

3. What *unit* is the partial pressure of oxygen measured in?

4. Preview the oxygen saturation curve and how it changes based on pH and temperature. What happens to oxygen when the curve shifts right? (**Use your own words**)

*Remember to show this pre-lab to the SSL instructor* **before entering** *the lab. Remember to show this prelab to the SSL instructor before entering lab for the lab activities. It will be marked in the SSL notebook for a grade."*

# 8

## RESPIRATORY PHYSIOLOGY
### IN-LAB ACTIVITIES

Name: _______________________    Section: __________    Date: _________

## LEARNING OBJECTIVES

1. Identify the respiratory muscles and describe their function.

2. Demonstrate the pressure and volume changes that occur during breathing.

3. Recognize the major lung volumes and capacities and how they change before and after exercise.

4. Have knowledge of the equation for carbon dioxide conversion into bicarbonate and understand what effect this would have on pH before and after exercise.

5. Recognize the oxygen-hemoglobin saturation curve and how this is affected by the partial pressure of oxygen and pH.

## PRE-LAB

*Before going to lab, you must complete the following:*

1. Read **Units 3.1, 3.2, & 3.3** lecture material located in Odigia Text (Breathing and Gas Exchange).

2. Read the **Pre-lab** and answer the questions that follow.

**Note:** You will spend **2 hours or so** in lab to complete the following activities. This amount of time allows you to complete the activities by using the torso model, using other models, and working with a lab partner.

*The following activities will be completed during lab:*

# INTRODUCTION: BREATHING AND GAS EXCHANGE

The respiratory system and the cardiovascular system are the two major organ systems of the body that are responsible for the exchange of gases such as oxygen and carbon dioxide. This exchange takes place using four main processes that include **ventilation, external respiration, gas transport, and internal respiration. Ventilation** or breathing allows air to be exchanged between the external environment and the exchange surfaces of the lungs. **External respiration** is the exchange of gases between the alveoli of the lungs and the pulmonary capillaries. **Gas transport** involves transport of gases within the blood. **Internal respiration** is the exchange of gases between the body cells and the blood in the systemic capillaries.

The process of ventilation or breathing consists of two major types of movements that include **inhalation** (or movement of air into the lungs) and **exhalation** (or movement of air out of lungs). Inspiration (or inhalation) and expiration (or exhalation) are dependent on the differences in pressure between the atmosphere and the lungs. **Boyle's Law** describes the relationship between volume and pressure in a gas at a constant temperature. Boyle discovered that the pressure of a gas is inversely proportional to its volume; if volume increases, pressure decreases. Likewise, if volume decreases, pressure increases.

**Watch the following video of Boyle's Law**

- https://www.youtube.com/watch?v=fH1H4iglW2A

In the human body, pressure changes are caused by volume changes in the rib cage and thoracic cavity. These volume changes are caused by the contraction of certain skeletal muscles. The difference in pressures drives **pulmonary ventilation** because air flows down a pressure gradient, that is, air flows from an area of higher pressure to an area of lower pressure. Air flows into the lungs largely due to a difference in pressure; atmospheric pressure is greater than **intra-alveolar pressure,** and intra-alveolar pressure is greater than **intrapleural pressure.** Air flows out of the lungs during expiration based on the same principle; pressure within the lungs becomes greater than the atmospheric pressure. Those respiratory muscles that involve expanding the thoracic cage and/or rib cage act to increase volume inside of the lungs and thus, to draw air into the lungs; therefore, they are called **inspiratory muscles.** The major inspiratory muscle includes the **diaphragm** but also the external intercostal muscles. Other skeletal muscles further help expand the rib cage but only during forceful breathing. These muscles include muscles of the rib cage and neck like the **sternocleidomastoid muscle, scalenes, and pectoralis minor.** Exhalation is normally a passive process in that after inspiration, the thoracic cage and rib cage go back to their normal position due to the compliance and elasticity of the lung and rib cage. During forceful exhalation however, other skeletal muscles contract that reduce the size of the rib cage and include major abdominal muscles as well as the **external intercostal muscles.**

## ACTIVITY 1: IDENTIFICATION AND FUNCTION OF RESPIRATORY MUSCLES AND SKELETAL STRUCTURES

*Refer back to lecture material Unit 3.1 and 3.2 as needed [Review muscles and bones.]*

1.  Using the models in lab, identify the following **respiratory muscles** and associated structures:

    a.  scalenes

    b.  sternocleidomastoid

    c.  pectoralis minor

    d.  intercostals (external and internal)

    e.  external oblique

    f.  internal oblique

    g.  transversus abdominis

    h.  rectus abdominis

    i.  diaphragm

        i.   central tendon

        ii.  aortic hiatus

        iii. inferior vena caval hiatus

        iv.  costophrenic angles

2.  Identify the following **skeletal structures** associated with respiratory system:

    a.  **pectoral girdle**

        i.   clavicle

        ii.  scapula

    b.  **sternum**

        i.   Manubrium

        ii.  body

        iii. xiphoid process

        iv.  manubrosternal junction (Angle of Lewis)

    c.  **ribs**

        i.   costal grooves (seen on the lungs surface)

## ACTIVITY 2: DEMONSTRATING PRESSURE AND VOLUME CHANGES IN THE RESPIRATORY SYSTEM

1. **Review** the video on Boyle's Law from the previous exercise if needed.

2. Find a **Bell Jar model** in the lab.

3. **Draw** a sketch of the Bell Jar model in the space provided.

4. **Label** the part of the Bell Jar that represents the following: lungs, trachea, bronchi, diaphragm, and thoracic cavity.

5. Move the part on the model that represents the **diaphragm** so that air is moving *into* the Bell Jar.

6. **Record** the shape of the diaphragm during inhalation in Table 8.1 below.

7. What happens to the volume inside of the Bell Jar? What happens to the pressure inside of the Bell Jar? Does it increase or decrease?

8. Now move the **diaphragm** so that air is moving *out of* the Bell Jar.

9. **Record** the shape of the diaphragm during exhalation in Table 8.1.

10. **Record** the muscles in humans that are involved in each type of movement.

**TABLE 8.1**

| MOVEMENT OF AIR | SHAPE OF DIAPHRAGM (FLATTENED OR DOMED UP) | VOLUME CHANGE (INCREASE OR DECREASE) | PRESSURE CHANGE (INCREASE OR DECREASE) | MUSCLES INVOLVED |
|---|---|---|---|---|
| Inhalation (inspiration) | | | | |
| Exhalation (expiration) | | | | |

## ACTIVITY 3: MEASURING LUNG VOLUMES AND CAPACITIES BEFORE AND AFTER EXERCISE

*Refer back to lecture material Unit 3.2 as needed [lung volumes and capacities].*

1. **Define** the following respiratory volumes.

   Tidal volume _______________________________________________

   Inspiratory reserve volume _______________________________________________

   Expiratory reserve volume _______________________________________________

   Residual volume _______________________________________________

2. **Describe** the following and show how they can be calculated from the respiratory volumes.

   Inspiratory capacity _______________________________________________

   Vital capacity _______________________________________________

   Functional residual capacity _______________________________________________

   Total lung capacity _______________________________________________

3. **Predict** how the following respiratory volumes or capacities might change after exercise before continuing with the activity.

   Tidal volume _______________________________________________

   Vital capacity _______________________________________________

   Functional residual capacity _______________________________________________

   Expiratory reserve volume _______________________________________________

*If using a spirometer that you only exhale from, also answer the following questions.*

- How might you calculate residual volume?

______________________________________________

______________________________________________

______________________________________________

______________________________________________

- How might you calculate inspiratory reserve volume?

______________________________________________

______________________________________________

______________________________________________

______________________________________________

4. You will be using a spirometer to measure the different lung volumes and lung capacities before and after exercise. In lab, follow the directions that came with the spirometer to measure tidal volume, expiratory reserve volume, and vital capacity *before* **exercise.**

5. **Record** this information in Table 8.2.

6. **Exercise for five minutes** and then use spirometer again to measure tidal volume, expiratory reserve volume, and vital capacity *after* **exercise.**

7. **Record** this information in Table 8.2.

**Lung volumes and capacities at rest and after exercise**

| TABLE 8.2 | | | | |
|---|---|---|---|---|
| LUNG VOLUME/ CAPACITY | DIRECTLY MEASURED OR CALCULATED? | AVERAGE VALUE *BEFORE* EXERCISE | AVERAGE VALUE *AFTER* EXERCISE | EQUATION FOR CALCULATION IF APPLICABLE |
| Tidal volume | | | | |
| Expiratory reserve volume | | | | |
| Inspiratory reserve volume | | | | |
| Residual volume | | | | |
| Vital capacity | | | | |
| Functional lung capacity | | | | |
| Inspiratory capacity | | | | |
| Total lung capacity | | | | |

# ACTIVITY 4: MEASURING EFFECT OF CARBON DIOXIDE LEVELS ON pH

1. **Read** about the transport of carbon dioxide in your online text. (*What Role does Carbon Dioxide Play*)

2. **List** the *three* main ways that carbon dioxide is transported in the blood.

---

---

---

3. Write the chemical equation that involves the reaction of carbon dioxide with water. What enzyme catalyzes this reaction?

_______________________________________________

_______________________________________________

_______________________________________________

**Predict** how the pH of a solution might change after exercise based on the chemical equation and your knowledge of respiratory volumes and capacities. Explain **why** you are predicting these results.

4. Fill up two paper cups with water. Label one before exercise and a second cup after exercise.

5. Using a pH meter or pH test strip, measure the pH of each labeled cup.

6. Do rigorous exercise for two-five minutes.

7. Using a straw, blow into the cup of water labeled *after* **exercise**.

8. Measure the pH of both cups of water and record the results in Table 8.3.

**pH levels from blowing into a cup before and after exercise**

| TABLE 8.3 | | |
| --- | --- | --- |
| **CUP LABEL** | **pH *BEFORE* EXERCISE** | **pH *AFTER* EXERCISE** |
| Before exercise | | |
| After exercise | | |

## DISCUSSION

1. Compare your results to your prediction; were they the same? How did they differ? Explain.

2. Why does pH of the blood change during exercise?

3. How does this change affect normal respiratory functions?

## ACTIVITY 5: PREDICTING HEMOGLOBIN-OXYGEN SATURATION USING THE HEMOGLOBIN-OXYGEN SATURATION CURVE

*Look at the following curve describing the relationship between the partial pressure of oxygen and the oxygen saturation of hemoglobin in the blood (Figure 8.1.) and **Answer Questions 1–5.***

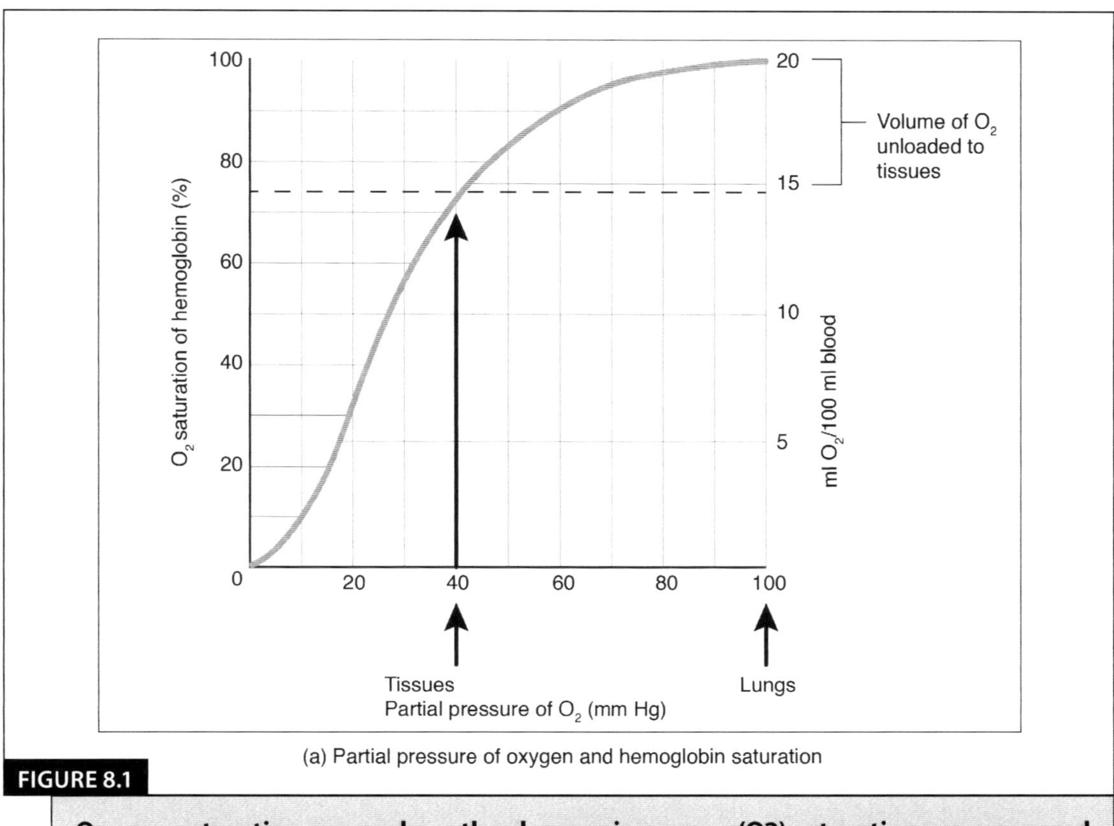


**FIGURE 8.1**

**Oxygen saturation curve where the changes in oxygen (O2) saturation are compared to the changes in the partial pressure of oxygen.** The volume of O2 per 100mL of blood is also displayed. Transport of Gases [https://cnx.org/contents/FPtK1zmh@8.108:3s_Xyutk@6/Transport-of-Gases] by OpenStax, (CC BY 4.0)

1. According to the graph, what is the percent saturation of the hemoglobin at the lungs?

2. Using the curve and the graph, estimate the partial pressure of oxygen that would allow the hemoglobin to be 80% saturated.

3. If the partial pressure of oxygen were to decrease to 30 mm Hg, what would be the approximate saturation of oxygen?

4. What is meant when the hemoglobin becomes more saturated? Less saturated? What would cause these changes?

5. Notice that between 40 mm Hg and 100 mm Hg of partial pressure of oxygen that the % saturation changes only from 75% to 100%. Why might a smaller change in % saturation of hemoglobin be beneficial despite a much larger change in partial pressure of oxygen?

Look at the following curves describing the relationship between partial pressure of oxygen, percent saturation of hemoglobin, and pH (Figure 8.2) and **answer questions 6-10.**

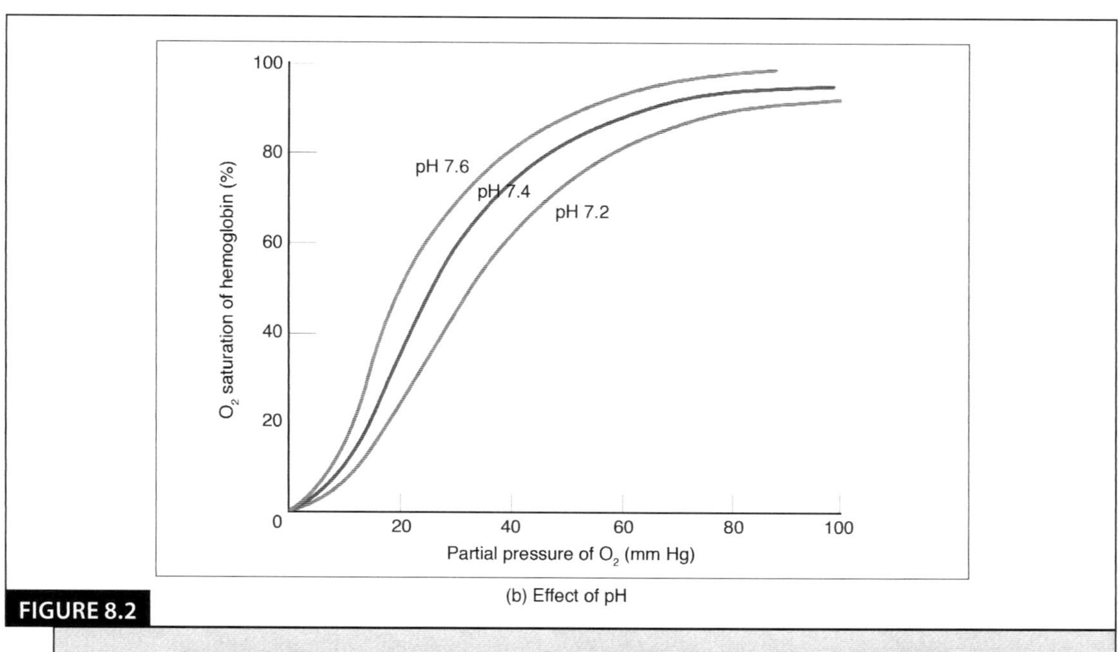


**FIGURE 8.2**

**Oxygen-hemoglobin saturation curve where the changes in oxygen (O2) are compared to changes of partial pressure of oxygen at three different pH** (7.2, 7.4, and 7.6)
Transport of Gases [https://cnx.org/contents/FPtK1zmh@8.108:3s_Xyutk@6/Transport-of-Gases] by OpenStax, (CC BY 4.0)

**6.** Look at 60 mm Hg partial pressure of oxygen to answer the following questions.

    **a.** What is the percent saturation at pH 7.4?_________________________________

    **b.** What is the percent saturation at pH 7.6?_________________________________

    **c.** What is the percent saturation at pH 7.2?_________________________________

**7.** At 60 mm Hg partial pressure of oxygen, did the saturation of hemoglobin increase or decrease at pH 7.2?  Did it increase or decrease at pH 7.6?

**8.** What might explain the difference in saturation of hemoglobin at the different pH? **Hint:** Review Activity 4 and think about or look up how carbon dioxide and acidity would affect hemoglobin saturation.

9. *Use the answers from the previous question and what you know about the effect of pH and carbon dioxide on oxygen saturation to answer the following.* When there is an increase in activity in the cells of the body, such as during muscle contraction, pH increases or decreases;, this results in the saturation of hemoglobin (increasing or decreasing), resulting in the offloading or release of (more or less) oxygen to the tissues. This causes a (right or left) shift in the oxygen-hemoglobin dissociation curve.

10. Alkalosis is a condition in which pH is excessively high. Based on the curves showing the relationship between pH and the dissociation curve, what effect would alkalosis have on the saturation of oxygen? Would this shift the oxygen dissociation curve toward the right or the left?

*Be sure to get your completed work checked off by a member of the lab staff and then keep this hand out for your review.*

# 9

# THE DIGESTIVE SYSTEM
## PRE-LAB

Name: _______________________  Section: __________  Date: ________

## LEARNING OBJECTIVES

1. Describe the basic processes of the digestive system.

2. Identify the components of the digestive system and their functions.

3. Examine the structure and function of the layers that compose the gastrointestinal tract.

4. Label and explain the five major folds of the peritoneum.

5. Describe each organ of the digestive system.

*Checklist to complete* **before entering** *the science skills lab (SSL):*

☐ Actively read this packet of information.

☐ Complete the charts, tables or labeling and answer questions using your own words.

☐ Complete the electronic digital pre-lab quiz on Bb by Sunday.

☐ Review the attached anatomy list and take it to lab with you. Jot down key descriptive identifying words that aid in your lab test preparations.

## ACTIVITY 1: DESCRIBE THE BASIC PROCESSES OF THE DIGESTIVE SYSTEM

*Read the content in your text and the introductory paragraphs below. Then, complete Activity 1 in lab.*

The processes of digestion include six activities: ingestion, propulsion, mechanical or physical digestion, chemical digestion, absorption, and defecation. The first of these processes, **ingestion,** refers to the entry of food into the alimentary canal through the mouth. There, the food is chewed and mixed with saliva, which contains enzymes that begin breaking down the carbohydrates in the food plus some lipid digestion via lingual lipase.

Digestion includes both mechanical and chemical processes. **Mechanical digestion** is a purely physical process that does not change the chemical nature of the food. Instead, it makes the food smaller to increase both surface area and mobility. It includes **mastication,** or chewing, as well as tongue movements that help break food into smaller bits and mix food with saliva. The mechanical churning of food in the stomach serves to further break it apart and expose more of its surface area to digestive juices, creating an acidic "soup" called **chyme. Segmentation,** which occurs mainly in the small intestine, consists of localized contractions continuously subdividing, breaking up, and mixing the contents.

In **chemical digestion,** starting in the mouth, digestive secretions break down complex food molecules into their chemical building blocks. The process is completed in the small intestine. Food that has been broken down is of no value to the body unless it enters the bloodstream and its nutrients are put to work. This occurs through the process of **absorption,** which takes place primarily within the small intestine. In **defecation,** the final step in digestion, undigested materials are removed from the body as feces.

On the figure below, list all of the processes of the digestive system that happen in that organ. Some may only have one answer; others may have many.

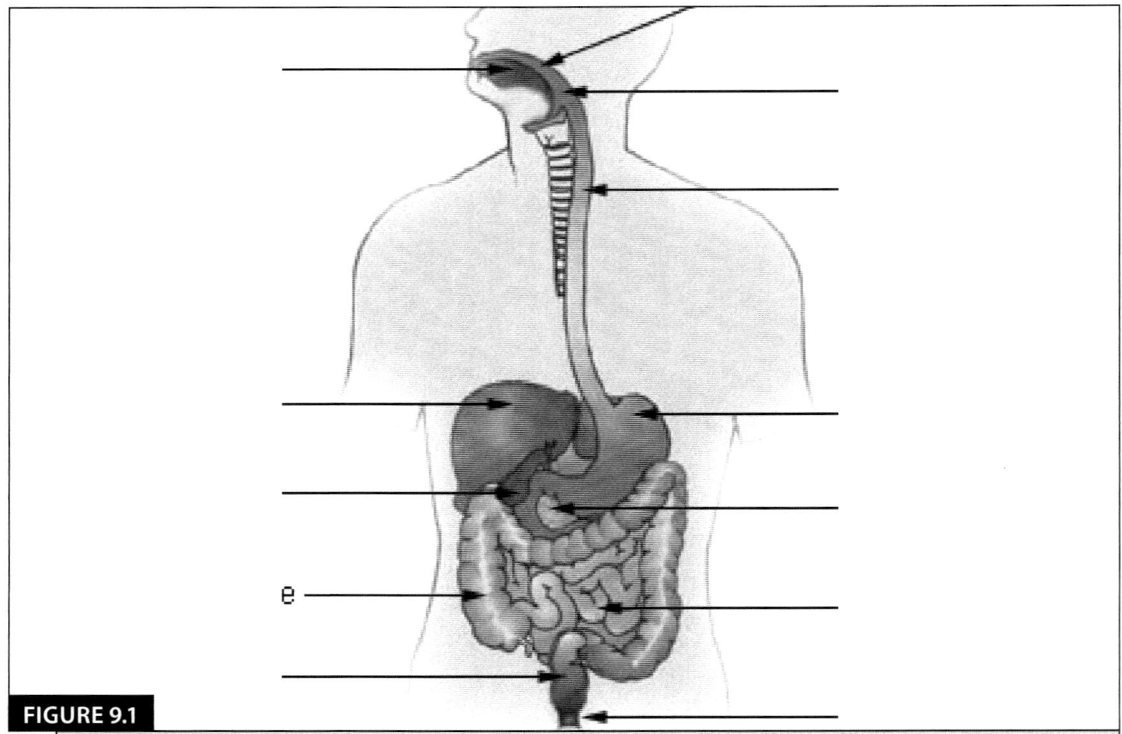

**FIGURE 9.1**

**The figure above is a schematic of the digestive system.** For each arrow, list all of the processes of the digestive system that occur there. Some may have only one process; others may have many. The six to choose from are ingestion, propulsion, mechanical or physical digestion, chemical digestion, absorption, and defecation. 

Public Domain

## ACTIVITY 2: IDENTIFY THE COMPONENTS OF THE DIGESTIVE SYSTEM AND THEIR FUNCTIONS

*Read the content in your text and the introductory paragraphs below. Then, complete Activity 2 in lab.*

Also called the gastrointestinal (GI) tract or gut, the **alimentary canal** is a one-way tube about 7.62 meters (25 feet) in length during life. The main function of the organs of the alimentary canal is to nourish the body. This tube begins at the mouth and terminates at the anus. Between those two points, the canal is modified as the pharynx, esophagus, stomach, and small and large intestines to fit the functional needs of the body. Both the mouth and anus are open to the external environment; thus, food and wastes within the alimentary canal are technically considered to be outside the body. Only through the process of absorption do the nutrients in food enter into and nourish the body's "inner space."

Each **accessory digestive organ** aids in the breakdown of food. Within the mouth, the teeth and tongue begin mechanical digestion, whereas the salivary glands begin chemical digestion. Once food products enter the small intestine, the gallbladder, liver, and pancreas release secretions—such as bile and enzymes—essential for digestion to continue. Together, these are called accessory organs because they sprout from the lining cells of the developing gut (mucosa) and augment its function; indeed, you could not live without their vital contributions, and many significant diseases result from their malfunction. Even after development is complete, they maintain a connection to the gut by way of ducts.

***Functions of each organ are described later in this Pre-Lab.***

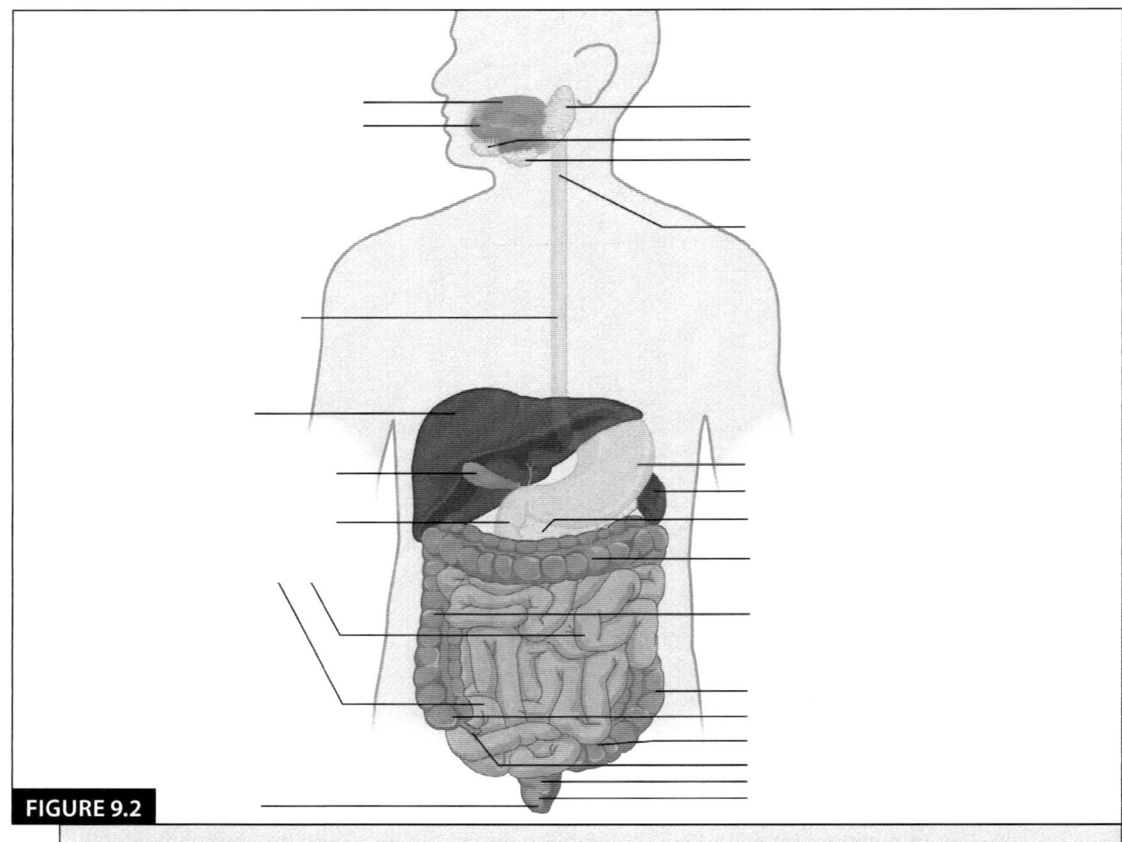

**FIGURE 9.2**

**On the drawing showing the organs of the digestive system above, label each component identified by a black line.** Overview of the Digestive System [https://cnx.org/contents/FPtK1zmh@8.108:IPvBytyq@4/Overview-of-the-Digestive-System] by OpenStax (CC BY 4.0)

## ACTIVITY 3: EXAMINE THE STRUCTURE AND FUNCTION OF THE LAYERS THAT COMPOSE THE GASTROINTESTINAL TRACT

*Read the content in your text and the introductory paragraphs below. Then, complete Activity 3 in lab.*

Throughout its length, the alimentary tract is composed of the same four tissue layers; the details of their structural arrangements vary to fit their specific functions. Starting from the lumen and moving outwards, these layers are the mucosa, submucosa, muscularis, and serosa, which is continuous with the mesentery.

The **mucosa** is referred to as a mucous membrane because mucous production is a characteristic feature of gut epithelium. The membrane consists of epithelium, which is in direct contact with ingested food, and the lamina propria, a layer of connective tissue analogous to the dermis. In addition, the mucosa has a thin, smooth muscle layer, called the muscularis mucosa (not to be confused with the muscularis layer).

As its name implies, the **submucosa** lies immediately beneath the mucosa. A broad layer of dense connective tissue, it connects the overlying mucosa to the underlying muscularis. It includes blood and lymphatic vessels (which transport absorbed nutrients) and a scattering of submucosal glands that release digestive secretions. Additionally, it serves as a conduit for a dense branching network of nerves, the submucosal plexus, which functions as described below.

The third layer of the alimentary canal is the **muscalaris** (also called the muscularis externa). The muscularis in the small intestine is made up of a double layer of smooth muscle: an inner circular layer and an outer longitudinal layer. The contractions of these layers promote mechanical digestion, expose more of the food to digestive chemicals, and move the food along the canal. The stomach is equipped for its churning function by the addition of a third layer, the oblique muscle. While the colon has two layers like the small intestine, its longitudinal layer is segregated into three narrow parallel bands, the tenia coli, which make it look like a series of pouches rather than a simple tube.

The **serosa** is the portion of the alimentary canal superficial to the muscularis. Present only in the region of the alimentary canal within the abdominal cavity, it consists of a layer of visceral peritoneum overlying a layer of loose connective tissue. Instead of serosa, the mouth, pharynx, and esophagus have a dense sheath of collagen fibers called the adventitia. These tissues serve to hold the alimentary canal in place near the ventral surface of the vertebral column.

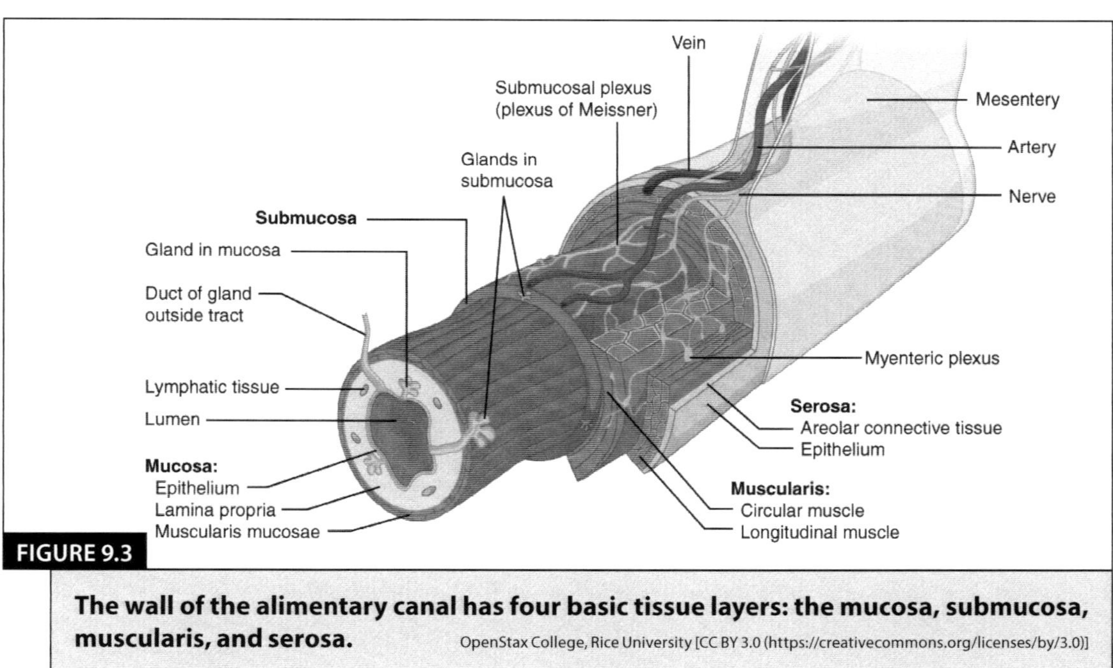


**FIGURE 9.3**

**The wall of the alimentary canal has four basic tissue layers: the mucosa, submucosa, muscularis, and serosa.** OpenStax College, Rice University [CC BY 3.0 (https://creativecommons.org/licenses/by/3.0)]

Fill in the table below. For each layer of the alimentary canal, give its component and that component's function.

| LAYER | COMPONENT | FUNCTION OF THE COMPONENT |
|---|---|---|
| Mucosa | | |
| | | |
| | | |
| Submucosa | | |
| Muscularis | | |
| | | |
| Serosa | | |

## ACTIVITY 4: LABEL AND EXPLAIN THE FIVE MAJOR FOLDS OF THE PERITONEUM

*Read the content in your text and the introductory paragraphs below. Then, complete Activity 4 in lab.*

The digestive organs within the abdominal cavity are held in place by the peritoneum, a broad serous membranous sac made up of squamous epithelial tissue surrounded by connective tissue. It is composed of two different regions: the parietal peritoneum, which lines the abdominal wall, and the visceral peritoneum, which envelops the abdominal organs. The peritoneal cavity is the space bounded by the visceral and parietal peritoneal surfaces. A few milliliters of watery fluid act as a lubricant to minimize friction between the serosal surfaces of the peritoneum.

The visceral peritoneum includes multiple large folds that envelop various abdominal organs, holding them to the dorsal surface of the body wall. Within these folds are blood vessels, lymphatic vessels, and nerves that innervate the organs with which they are in contact, supplying their adjacent organs. Note that during fetal development, certain digestive structures, including the first portion of the small intestine (called the duodenum), the pancreas, and portions of the large intestine (the ascending and descending colon and the rectum) remain completely or partially posterior to the peritoneum. Thus, the location of these organs is described as **retroperitoneal.**

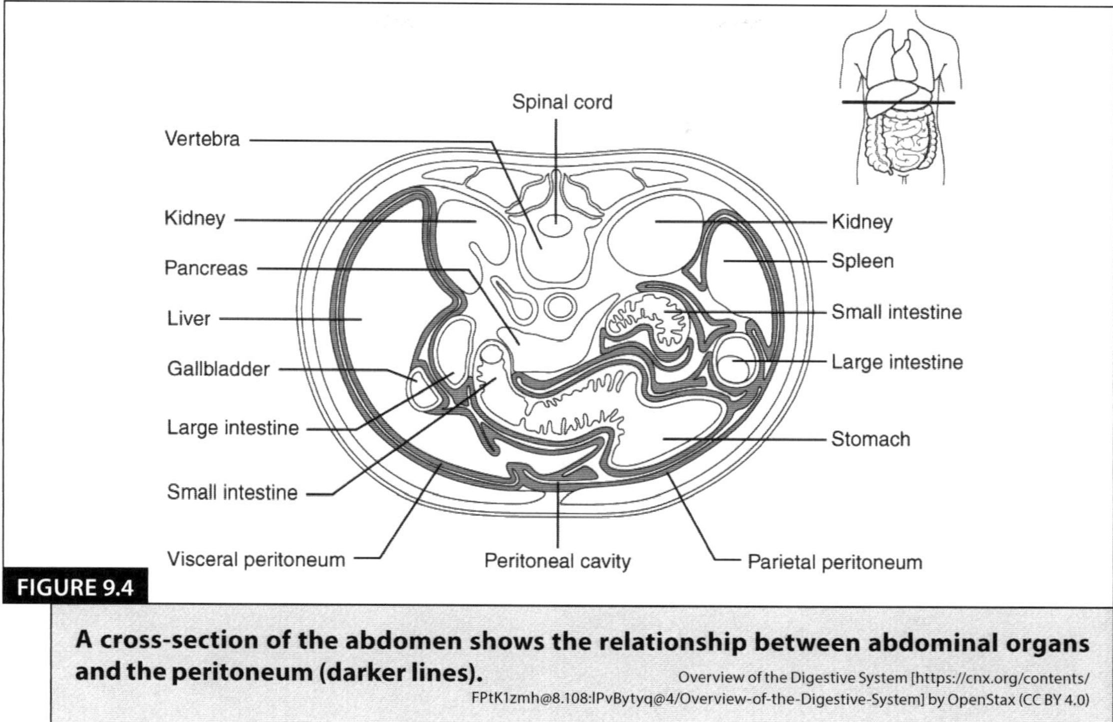



**FIGURE 9.4**

**A cross-section of the abdomen shows the relationship between abdominal organs and the peritoneum (darker lines).** Overview of the Digestive System [https://cnx.org/contents/ FPtK1zmh@8.108:IPvBytyq@4/Overview-of-the-Digestive-System] by OpenStax (CC BY 4.0)

Fill in the table below. For each peritoneal fold, describe it in the space provided.

| FOLD | DESCRIPTION |
| --- | --- |
| Greater omentum | |
| Falciform ligament | |
| Lesser omentum | |
| Mesentery | |
| Mesocolon | |

# ACTIVITY 5: DESCRIBE EACH ORGAN OF THE DIGESTIVE SYSTEM

*Read the content in your text and the introductory paragraphs below. Then, complete Activity 5 in lab.*

## THE MOUTH

The cheeks, tongue, and palate frame the mouth, which is also called the **oral cavity** (or buccal cavity). At the entrance to the mouth are the lips, or **labia** (singular = labium). Their outer covering is skin, which transitions to a mucous membrane in the mouth proper. The **labial frenulum** is a midline fold of mucous membrane that attaches the inner surface of each lip to the gum. The cheeks make up the oral cavity's sidewalls. Moving farther into the mouth, the opening between the oral cavity and throat (oropharynx) is called the **fauces** (like the kitchen "faucet").

The roof of your mouth is called the palate. The anterior region of the palate serves as a wall (or septum) between the oral and nasal cavities as well as a rigid shelf against which the tongue can push food. It is created by the maxillary and palatine bones of the skull and, given its bony structure, is known as the **hard palate.** If you run your tongue along the roof of your mouth, you'll notice that the hard palate ends in the posterior oral cavity, and the tissue becomes fleshier. This part of the palate, known as the **soft palate,** is composed mainly of skeletal muscle.

## THE PHARYNX

The **pharynx** (throat) is involved in both digestion and respiration. It receives food and air from the mouth and air from the nasal cavities. When food enters the pharynx, involuntary muscle contractions close off the air passageways. A short tube of skeletal muscle lined with a mucous membrane, the pharynx runs from the posterior oral and nasal cavities to the opening of the esophagus and larynx. It has three subdivisions. The most superior, the **nasopharynx,** is involved only in breathing and speech. The other two subdivisions, the **oropharynx** and the **laryngopharynx,** are used for both breathing and digestion. The oropharynx begins inferior to the nasopharynx and is continuous below with the laryngopharynx.

## THE SALIVARY GLANDS

Many small **salivary glands** are housed within the mucous membranes of the mouth and tongue. Secretion increases when you eat because saliva is essential to moisten food and initiate the chemical breakdown of carbohydrates. Outside the oral mucosa are three pairs of major salivary glands, which secrete the majority of saliva into ducts that open into the mouth. The **submandibular glands,** which are in the floor of the mouth, secrete saliva into the mouth through the submandibular ducts. The **sublingual glands,** which lie below the tongue, use the lesser sublingual ducts to secrete saliva into the oral cavity. The **parotid glands** lie between the skin and the masseter muscle near the ears.

## THE ESOPHAGUS

The **esophagus** is a muscular tube that connects the pharynx to the stomach. It is approximately 25.4 cm (10 in) in length, located posterior to the trachea, and remains in a collapsed form when not engaged in swallowing. The esophagus runs a mainly straight route through the mediastinum of the thorax. The **upper esophageal sphincter,** which is continuous with the inferior pharyngeal constrictor, controls the movement of food from the pharynx into the esophagus. Food passes from the esophagus into the stomach at the **lower esophageal sphincter.** The lower esophageal sphincter relaxes to let food pass into the stomach and then contracts to prevent stomach acids from backing up into the esophagus.

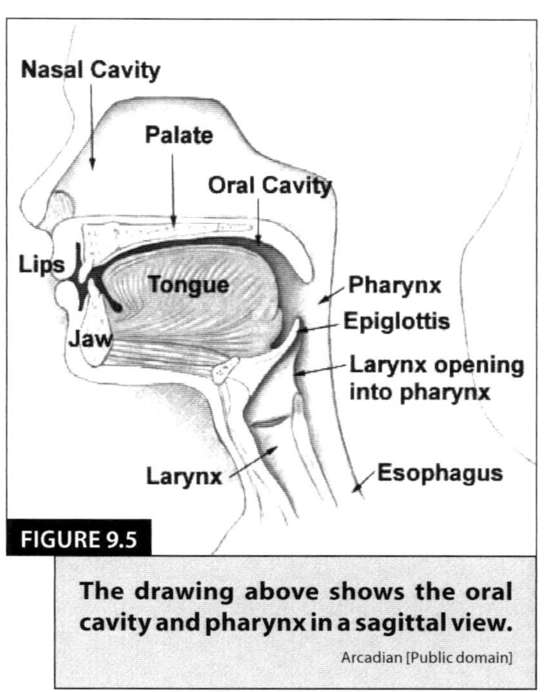


**FIGURE 9.5**

**The drawing above shows the oral cavity and pharynx in a sagittal view.**

Arcadian [Public domain]

## THE STOMACH

Although a minimal amount of carbohydrate digestion occurs in the mouth, chemical digestion really gets underway in the stomach. There are four main regions in the **stomach:** the cardia, fundus, body, and pylorus. The **cardia** (or cardiac region) is the point where the esophagus connects to the stomach and through which food passes into the stomach. Located inferior to the diaphragm, above and to the left of the cardia, is the dome-shaped **fundus.** Below the fundus is the **body,** the main part of the stomach. The funnel-shaped **pylorus** connects the stomach to the duodenum. The wider end of the funnel, the **pyloric antrum,** connects to the body of the stomach. The narrower end is called the **pyloric canal,** which connects to the duodenum. The smooth muscle **pyloric sphincter** is located at this latter point of connection and controls stomach emptying. In the absence of food, the stomach deflates inward, and its mucosa and submucosa fall into a large fold called a **rugae.**

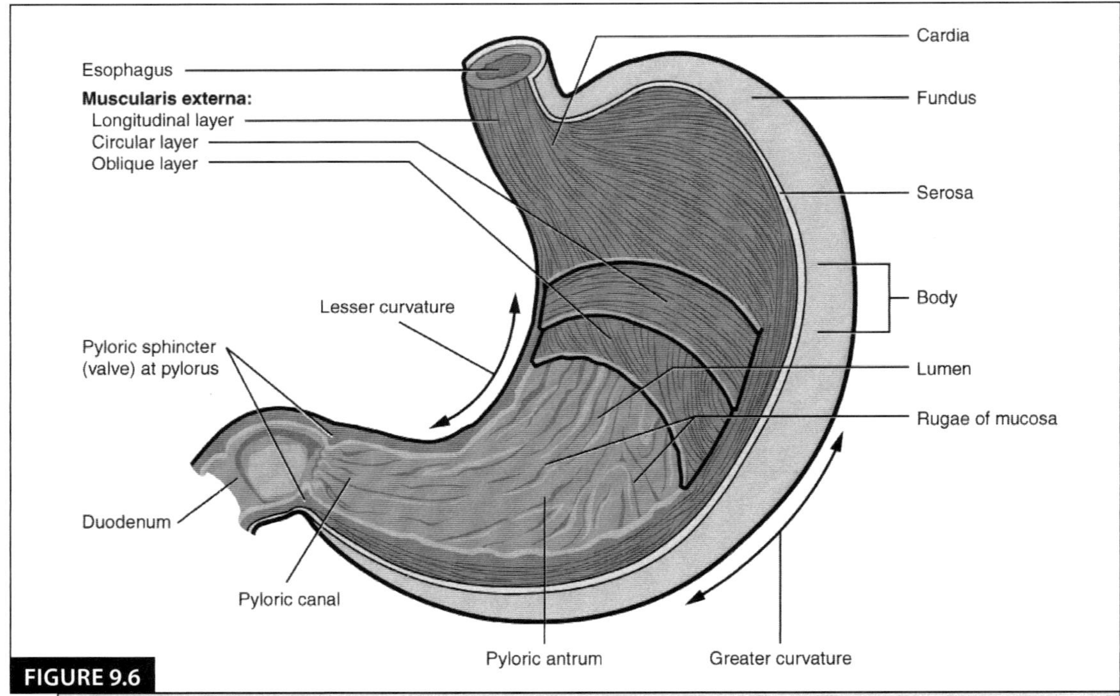


**FIGURE 9.6**

**The stomach has four major regions: the cardia, fundus, body, and pylorus.** The addition of an inner oblique smooth muscle layer gives the muscularis the ability to vigorously churn and mix food. The Stomcach [https://cnx.org/contents/FPtK1zmh@8.108:O9dvCxUQ@3/The-Stomach] by OpenStax (CC BY 4.0)

## THE SMALL INTESTINE

Chyme released from the stomach enters the **small intestine,** which is the primary digestive organ in the body. Not only is this where most digestion occurs, it is also where practically all absorption occurs. The longest part of the alimentary canal, the small intestine, is about 3.05 meters (10 feet) long in a living person. The coiled tube of the small intestine is subdivided into three regions. The shortest region is the 25.4-cm (10-in) **duodenum,** which begins at the pyloric sphincter. The **jejunum** is about 0.9 meters (3 feet) long (in life) and runs from the duodenum to the ileum. The **ileum** is the longest part of the small intestine, measuring about 1.8 meters (6 feet) in length. The ileum joins the cecum, the first portion of the large intestine, at the **ileocecal sphincter** (or valve).

Also called a plica circulare, a **circular fold** is a deep ridge in the mucosa and submucosa. Beginning near the proximal part of the duodenum and ending near the middle of the ileum, these folds facilitate absorption. Within the circular folds are small (0.5–1 mm long) hairlike vascularized projections called **villi** (singular = villus) that give the mucosa a furry texture. **Microvilli** (singular = microvillus) are much smaller (1 μm) than villi. They are cylindrical apical surface extensions of the plasma membrane of the mucosa's epithelial cells. Although their small size makes it difficult to see each microvillus, their combined microscopic appearance suggests a mass of bristles, which is termed the **brush border.** Fixed to the surface of the microvilli, membranes are enzymes that finish digesting carbohydrates and proteins.

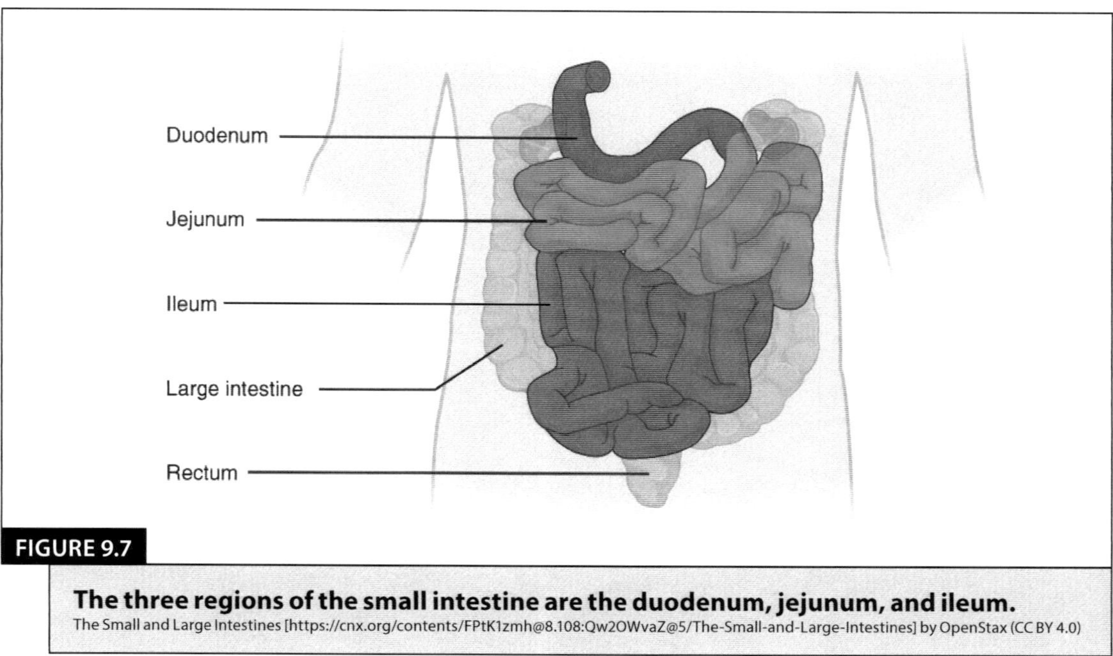


**FIGURE 9.7**

**The three regions of the small intestine are the duodenum, jejunum, and ileum.**
The Small and Large Intestines [https://cnx.org/contents/FPtK1zmh@8.108:Qw2OWvaZ@5/The-Small-and-Large-Intestines] by OpenStax (CC BY 4.0)

## THE LARGE INTESTINE

The **large intestine** is the terminal part of the alimentary canal. The primary function of this organ is to finish absorption of nutrients and water, synthesize certain vitamins, form feces, and eliminate feces from the body. The first part of the large intestine is the **cecum,** a sac-like structure that is suspended inferior to the ileocecal valve. It is about 6 cm (2.4 in) long, receives the contents of the ileum, and continues the absorption of water and salts. The **appendix** (or vermiform appendix) is a winding tube that attaches to the cecum; this organ is generally considered vestigial. The cecum blends seamlessly with the **colon.** Upon entering the colon, the food residue first travels up the **ascending colon** on the right side of the abdomen. At the inferior surface of the liver, the colon bends to form the **right colic flexure** (hepatic flexure) and becomes the **transverse colon.** The region defined as hindgut begins with the last third of the transverse colon and continues on. Food residue passing through the transverse colon travels across to the left side of the abdomen, where the colon angles sharply immediately inferior to the spleen, at the **left colic flexure** (splenic flexure). From there, food residue passes through the **descending colon,** which runs down the left side of the posterior abdominal wall. After entering the pelvis inferiorly, it becomes the s-shaped **sigmoid colon,** which extends medially to the midline.

Food residue leaving the sigmoid colon enters the **rectum** in the pelvis, near the third sacral vertebra. The final 20.3 cm (8 in) of the alimentary canal, the rectum extends anterior to the sacrum and coccyx. Lastly, the **anal canal,** which is located in the perineum, is completely outside of the abdominopelvic cavity. This 3.8–5 cm (1.5–2 in) long structure opens to the exterior of the body at the anus. The anal canal includes two sphincters. The **internal anal sphincter** is made of smooth muscle, and its contractions are involuntary. The **external anal sphincter** is made of skeletal muscle which is under voluntary control. Except when defecating, both usually remain closed.

Three features are unique to the large intestine: teniae coli, haustra, and epiploic append-ages. The **teniae coli** are three bands of smooth muscle that make up the longitudinal muscle layer of the muscularis of the large intestine, except at its terminal end. Tonic contractions of the teniae coli bunch up the colon into a succession of pouches called **haustra** (singular = hostrum), which are responsible for the wrinkled appearance of the colon. Attached to the teniae coli are small, fat-filled sacs of visceral peritoneum called **epiploic appendages.**

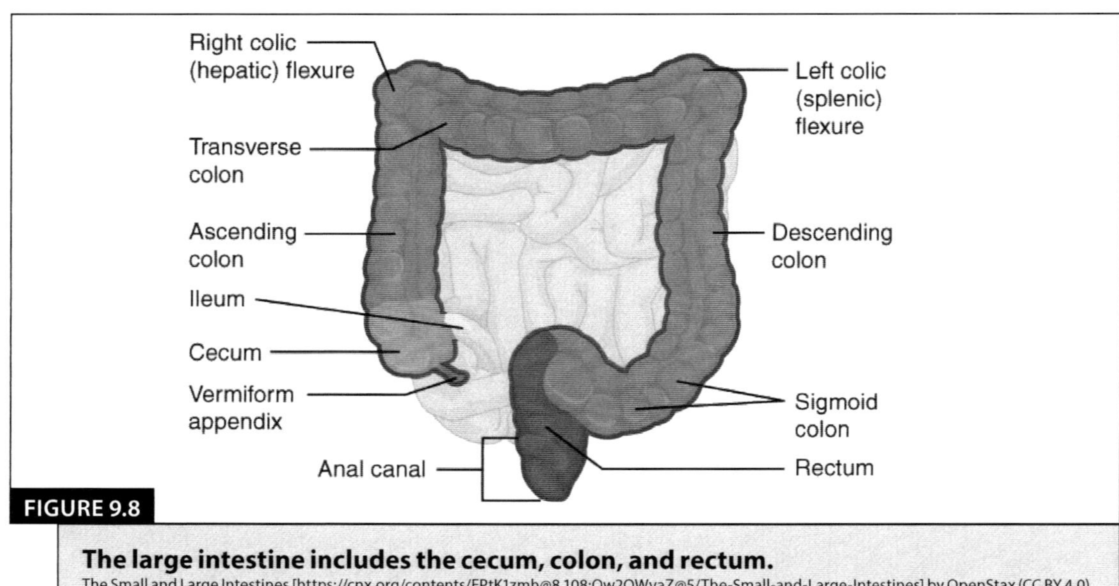


**FIGURE 9.8**

**The large intestine includes the cecum, colon, and rectum.**
The Small and Large Intestines [https://cnx.org/contents/FPtK1zmh@8.108:Qw2OWvaZ@5/The-Small-and-Large-Intestines] by OpenStax (CC BY 4.0)

Match each organ with the function.

1. Mouth       ________ Most digestion and absorption occurs here

2. Salivary glands       ________ Passageway for food and air

3. Pharynx       ________ Finish nutrient absorption and form feces

4. Esophagus       ________ Begins chemical digestion of food

5. Small intestine       ________ Contains fauces

6. Large intestine       ________ Remains collapsed unless it is in use

# ACCESSORY ORGANS: THE LIVER, PANCREAS, AND GALLBLADDER

## THE LIVER

The **liver** is the largest gland in the body, weighing about three pounds in an adult. The digestive role of the liver is to produce bile and export it to the duodenum. The liver is divided into two primary lobes: a large right lobe and a much smaller left lobe. In the right lobe, there is an inferior **quadrate lobe** and a posterior **caudate lobe.** The liver is connected to the abdominal wall and diaphragm by five peritoneal folds referred to as ligaments. We have already discussed the falciform ligament earlier. The **hepatic artery** delivers oxygenated blood from the heart to the liver. The **hepatic portal vein** delivers partially deoxygenated blood containing nutrients absorbed from the small intestine and actually supplies more oxygen to the liver than do the much smaller hepatic arteries. In addition to nutrients, drugs and toxins are also absorbed.

A **hepatocyte** is the liver's main cell type; these cells play a role in a wide variety of secretory, metabolic, and endocrine functions. Plates of hepatocytes called hepatic laminae radiate outward from the portal vein in each **hepatic lobule.** The **portal triad** is a distinctive arrangement around the perimeter of hepatic lobules, consisting of three basic structures: a bile duct, a hepatic artery branch, and a hepatic portal vein branch. Between adjacent hepatocytes, grooves in the cell membranes provide room for each bile canaliculus (plural = canaliculi). These small ducts accumulate the bile produced by hepatocytes. From here, bile flows into bile ducts that unite to form the larger right and left hepatic ducts, which themselves merge and exit the liver as the **common hepatic duct.** This duct then joins with the cystic duct from the gallbladder, forming the **common bile duct** through which bile flows into the small intestine.

## THE PANCREAS

The soft, oblong, glandular **pancreas** lies transversely in the retroperitoneum behind the stomach. The exocrine part of the pancreas arises as little grape-like cell clusters, each called an **acinus** (plural = acini), located at the terminal ends of pancreatic ducts. These acinar cells secrete enzyme-rich **pancreatic juice,** which contains digestive enzymes and bicarbonate ions, into two dominant ducts. The larger duct fuses with the common bile duct (carrying bile from the liver and gallbladder) just before entering the duodenum via a common opening (the **hepatopancreatic ampulla**). The second and smaller pancreatic duct, the **accessory duct** (duct of Santorini), runs from the pancreas directly into the duodenum, approximately 1 inch above the hepatopancreatic ampulla.

## THE GALLBLADDER

The **gallbladder** is 8–10 cm (~3–4 in) long and is nested in a shallow area on the posterior aspect of the right lobe of the liver. This muscular sac stores, concentrates, and, when stimulated, propels the bile into the duodenum via the common bile duct.

Label the figure below.

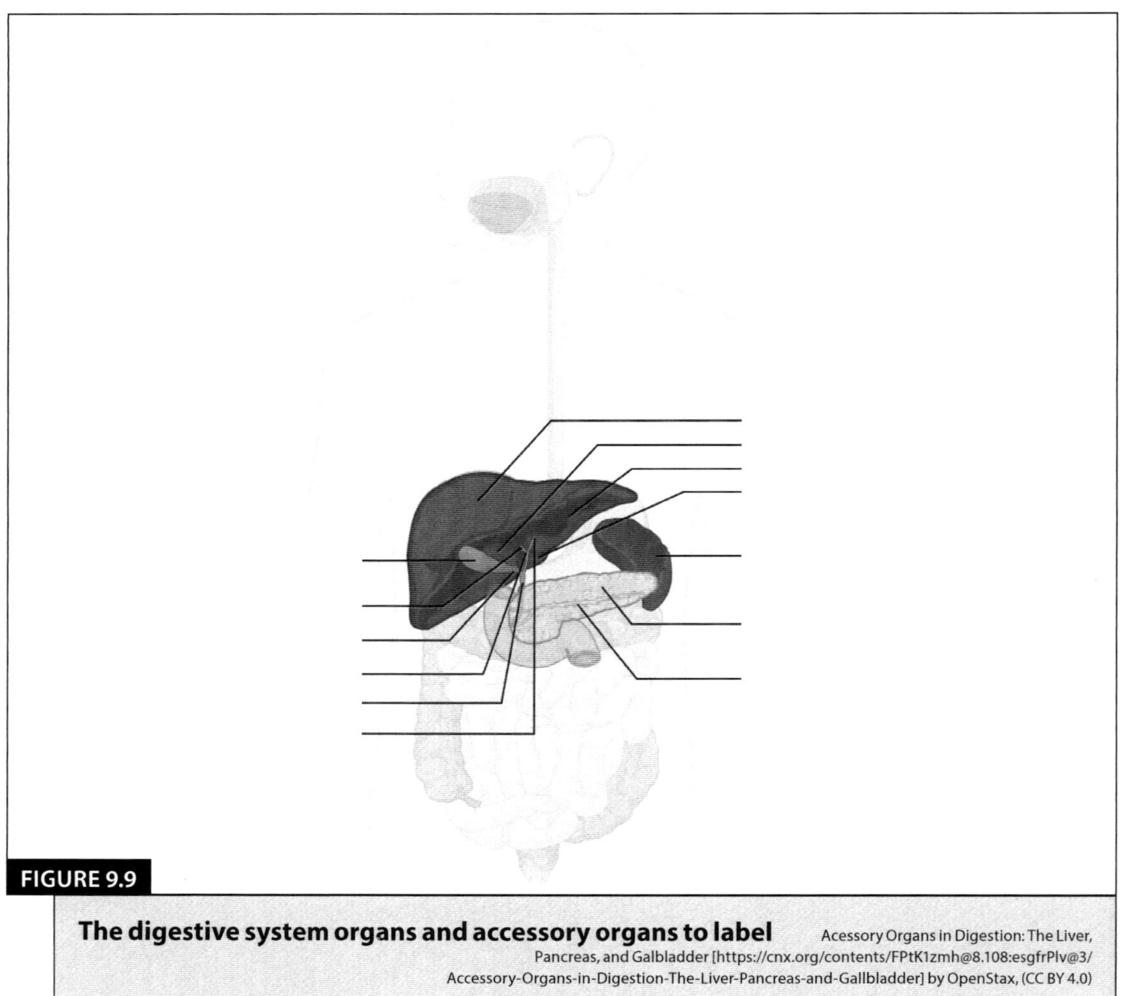

**FIGURE 9.9**

**The digestive system organs and accessory organs to label** Acessory Organs in Digestion: The Liver, Pancreas, and Galbladder [https://cnx.org/contents/FPtK1zmh@8.108:esgfrPlv@3/ Accessory-Organs-in-Digestion-The-Liver-Pancreas-and-Gallbladder] by OpenStax, (CC BY 4.0)

*Checklist to complete* **before entering** *the science skills lab (SSL):*

☐ Actively read this packet of information.

☐ Complete the charts, tables or labeling and answer questions using your own words.

☐ Complete the electronic digital pre-lab quiz on Bb by Sunday.

☐ Review the attached anatomy list and take it to lab with you. Jot down key descriptive identifying words that aid in your lab test preparations.

## ANATOMY LIST

1. Topographical landmarks. (You must be able to describe which abdominal organs (and parts of organs) lie in each quadrant and region.)

    a. Abdominal cavity quadrants (clinical)

    b. Abdominal cavity regions (anatomical)

2. Identify the following structures on models and diagrams:

    a. Salivary glands

        i. Parotid glands

            1. Parotid duct and opening

        ii. Submandibular glands

            1. Submandibular ducts and openings

        iii. Sublingual glands

            1. Sublingual ducts

    b. Esophagus

        i. Lower esophageal sphincter

    c. Stomach

        i. Cardia

        ii. Fundus

        iii. Body

        iv. Pylorus (pyloric canal)

        v. Pyloric sphincter

        vi. Greater curvature

        vii. Greater omentum

        viii. Lesser curvature

        ix. Lesser omentum

        x. Muscle layers

            1. Longitudinal

            2. Circular

            3. Oblique

        xi. Rugae

        xii. Gastric mucosa

    **d.** Small Intestine
- **i.** Duodenum
- **ii.** Plicae circulare
- **iii.** Duodenal papilla
  1. hepatopancreatic ampulla
- **iv.** Jejunum
- **v.** Ileum
- **vi.** Mesentery
- **vii.** Superior mesenteric artery and vein
- **viii.** Inferior mesenteric artery and vein
- **ix.** Ileocecal valve

    **e.** Colon
- **i.** Haustra
- **ii.** Taenia coli
- **iii.** Epiploic appendages
- **iv.** Cecum
- **v.** Vermiform appendix
- **vi.** Ascending colon
- **vii.** Hepatic flexure
- **viii.** Transverse colon
- **ix.** Splenic flexure
- **x.** Descending colon
- **xi.** Sigmoid colon
- **xii.** Rectum
  1. Transverse rectal folds
  2. Rectal venous plexus
- **xiii.** Anus
  1. Internal anal sphincter
  2. External anal sphincter
  3. Levator ani muscles

    **f.** Liver
- **i.** Surfaces
  1. Anterior
  2. Posterior

    **ii.** Lobes

        1. Right

        2. Left

        3. Caudate

        4. Quadrate

    **iii.** Gall bladder

        1. Fundus

        2. Body

        3. Neck

    **iv.** Porta hepatis

        1. Hepatic artery

        2. Cystic artery

        3. Hepatic portal vein

        4. Hepatic ducts

        5. Cystic duct

        6. Common bile duct

    **v.** Hepatic veins

    **vi.** Inferior vena cava

    **vii.** Coronary ligament

    **viii.** Bare area

    **ix.** Falciform ligament

    **x.** Round ligament (ligamentum teres)

**g.** Pancreas

    **i.** Head

    **ii.** Body

    **iii.** Tail

    **iv.** Primary pancreatic duct

    **v.** Accessory pancreatic duct

3. Identify the microscopic structures indicated on the following slides:
   a. Esophagus—stomach junction
      i. Mucosa of esophagus—stratified squamous epithelium
      ii. Mucosa of stomach—simple columnar epithelium
      iii. Submucosa of esophagus
      iv. Esophageal glands
   b. Stomach
      i. Mucosa—simple columnar epithelium
      ii. Goblet cells
      iii. Gastric pits
      iv. Chief cells
      v. Parietal cells
      vi. Gastric glands
   c. Duodenum
      i. Plicae circularis
      ii. Intestinal crypts
      iii. Mucosa—simple columnar epithelium
      iv. Muscularis mucosa
      v. Submucosa
      vi. Brunner's glands
      vii. Villus
      viii. Muscularis externa
         1. Circular layer
         2. Longitudinal layer
      ix. Serosa
   d. Ileum
      i. Villi
      ii. Peyer's patches
      iii. Identify all other layers as listed above
   e. Colon
      i. Mucosa
      ii. Goblet cells
      iii. Taenia coli—longitudinal muscle layer
      iv. Identify all other layers as listed above

    **f.** Liver

        **i.** Hepatocytes

        **ii.** Kupffer cells

        **iii.** Hepatic triad

            1. Hepatic artery

            2. Hepatic portal vein

            3. Bile ductile

        **iv.** Lobule

        **v.** Sinusoids

        **vi.** Central vein

        **vii.** Interlobular septum

  **g.** Pancreas

        **i.** Pancreatic acini (acinar cells)

        **ii.** Islets of Langerhans (pancreatic islets)

        **iii.** Pancreatic ducts

*Remember to show this pre-lab to the SSL instructor* **before entering** *the lab. It will be marked off in the SSL book for grade credit towards your lab test.*

# 9

# THE DIGESTIVE SYSTEM
## IN-LAB ACTIVITIES

Name: _______________________     Section: __________     Date: _________

## ACTIVITY 1: DESCRIBE THE BASIC PROCESSES OF THE DIGESTIVE SYSTEM

*You will need a Carolina BioKit: Digestion, lab protocol, and materials (if available) to complete this exercise.*

1. Complete the table below by giving the definitions for each process of the digestive system.

| PROCESS | DEFINITION |
| --- | --- |
| Ingestion | |
| Mechanical Digestion | |
| Chemical Digestion | |
| Absorption | |
| Defecation | |

2. On the model below, indicate where ingestion, mechanical digestion, chemical digestion, absorption, and defecation occur (designate with an "I," "MD," "CD," "A," and "D" respectively).

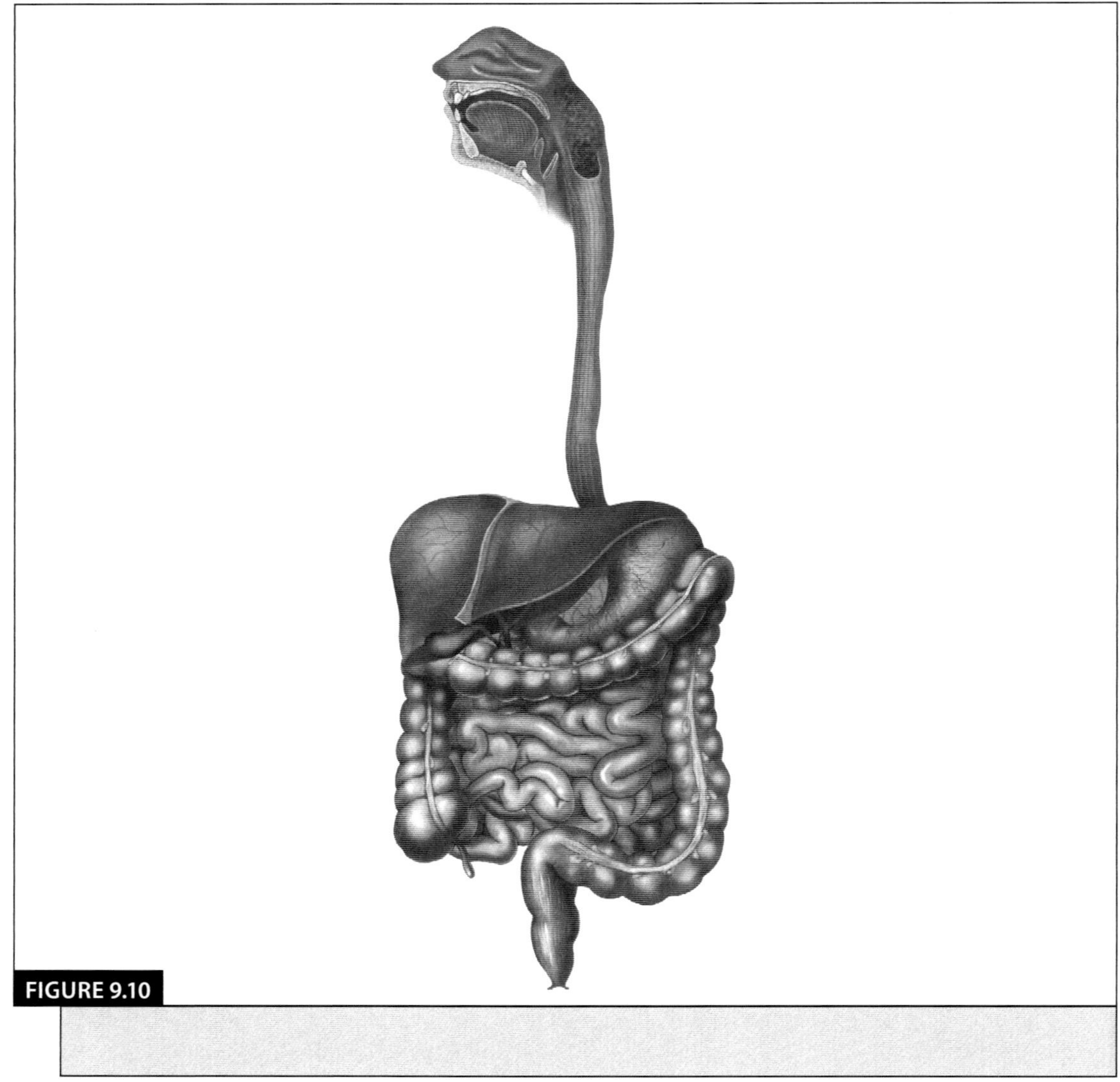

**FIGURE 9.10**

3. What is the difference between peristalsis and segmentation?

4. Perform the enzymatic digestion experiment following Carolina BioKit: Digestion instructions.

   **Note:** Lab kit can be found here: http://www.carolina.com/cellular-physiology-enzymes/carolina-biokits-digestion/202340.pr?question=

## ACTIVITY 2: IDENTIFY THE COMPONENTS OF THE DIGESTIVE SYSTEM AND THEIR FUNCTIONS

*You will need a full torso model (or the digestive board model if available) to complete this exercise.*

1. Obtain a full torso model (or the digestive board model if available) and locate the following components of the digestive system: the anus, esophagus, gallbladder, large intestine, liver, oral cavity, pancreas, pharynx, salivary glands, small intestine, and stomach.

2. Label the following components of the model below: the anus, esophagus, gallbladder, large intestine, liver, oral cavity, pancreas, pharynx, salivary glands, small intestine, and stomach.

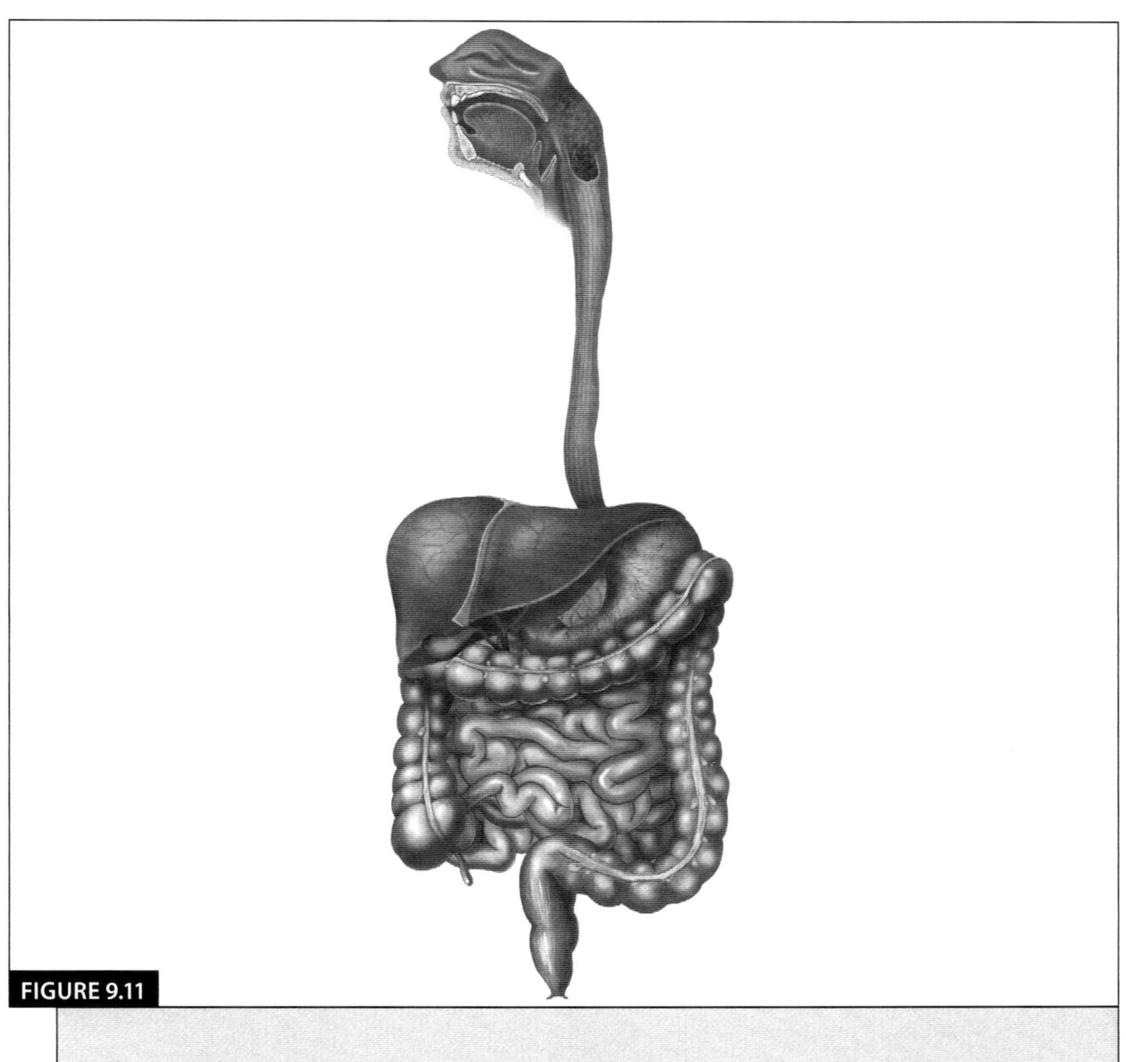

**FIGURE 9.11**

3.  Complete the chart below by giving the full function of each organ of the digestive system.

| | ORGAN | FUNCTION |
|---|---|---|
| Alimentary Canal Organs | | |
| | | |
| | | |
| | | |
| | | |
| | | |
| Accessory Organs | | |
| | | |
| | | |
| | | |
| | | |
| | | |

## ACTIVITY 3: EXAMINE THE STRUCTURE AND FUNCTION OF THE LAYERS THAT COMPOSE THE GASTROINTESTINAL TRACT

*You will need a villi model (if available) to complete this lab.*

1. In the space provided, label the numbered diagram below.

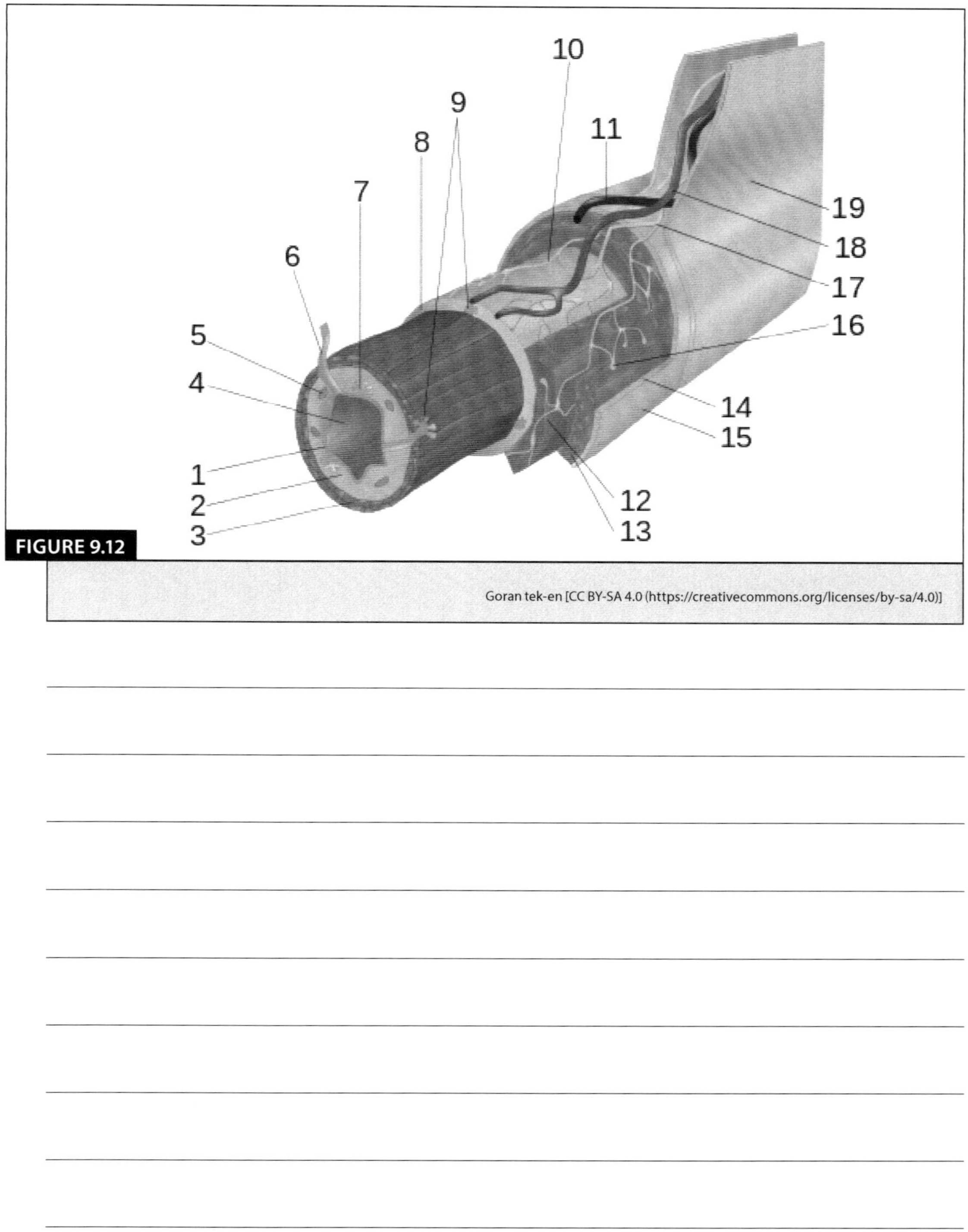

**FIGURE 9.12**

Goran tek-en [CC BY-SA 4.0 (https://creativecommons.org/licenses/by-sa/4.0)]

## ACTIVITY 4: LABEL AND EXPLAIN THE FIVE MAJOR FOLDS OF THE PERITONEUM

*You will need a full torso model, digestive board model, and loose digestive system models (if available) to complete this exercise.*

1.  Obtain the models listed above and locate the major folds of the peritoneum: mesentery, mesocolon, falciform ligament, greater omentum, and lesser omentum.

2.  Why are these folds essential to proper digestive system function?

3.  Explain the difference between the visceral and parietal peritoneum. What is the purpose of the fluid between them?

## ACTIVITY 5: DESCRIBE EACH ORGAN OF THE DIGESTIVE SYSTEM

*You will need a full torso model, digestive board model, loose digestive system models, and histology slides of the digestive system (if available) to complete this exercise. **Make sure to identify the structure on the model in front of you** and then **label it on the picture provided. Interacting with the models is an important kinesthetic component to lab.***

1. Label all of the following components of the oral cavity and pharynx on the model below: **cheek, esophagus, fauces, gingivae, hard palate, inferior labial frenulum, inferior lip, lingual tonsil, palatine tonsil, pharyngeal tonsil, soft palate, tongue, uvula,** and **vestibule.**

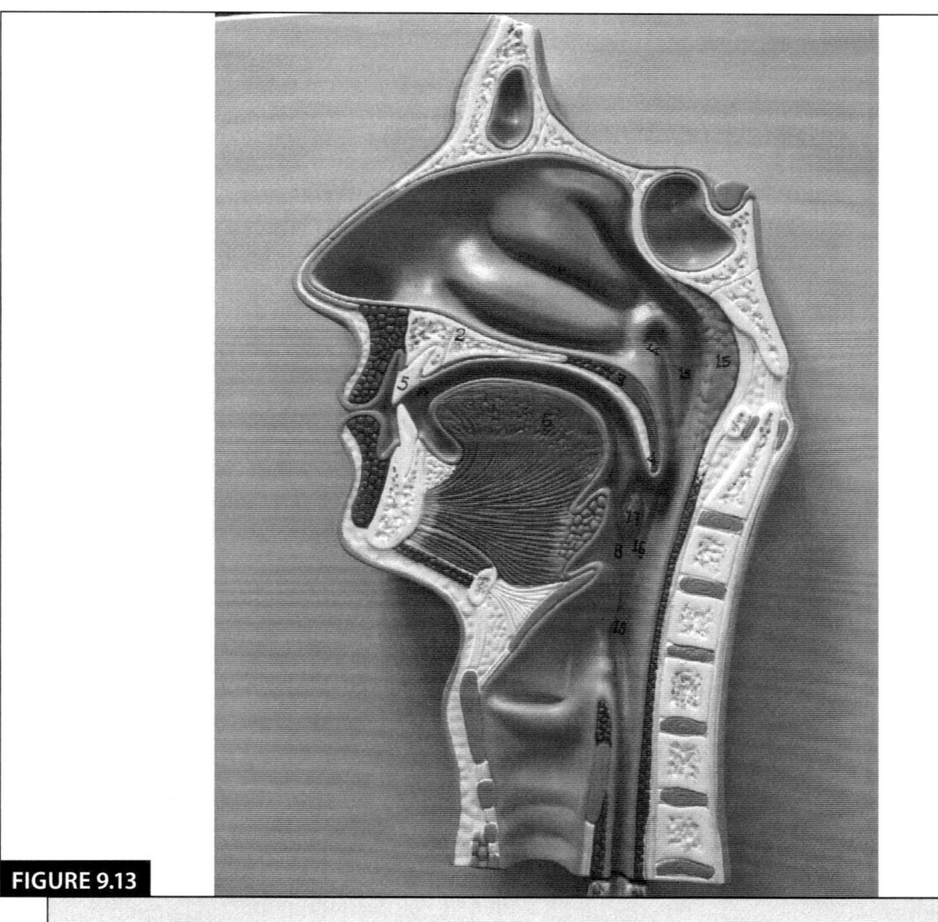

**FIGURE 9.13**

2. Label the three salivary glands on the model below. [See the model legend in the lab or pictures in your online text.]

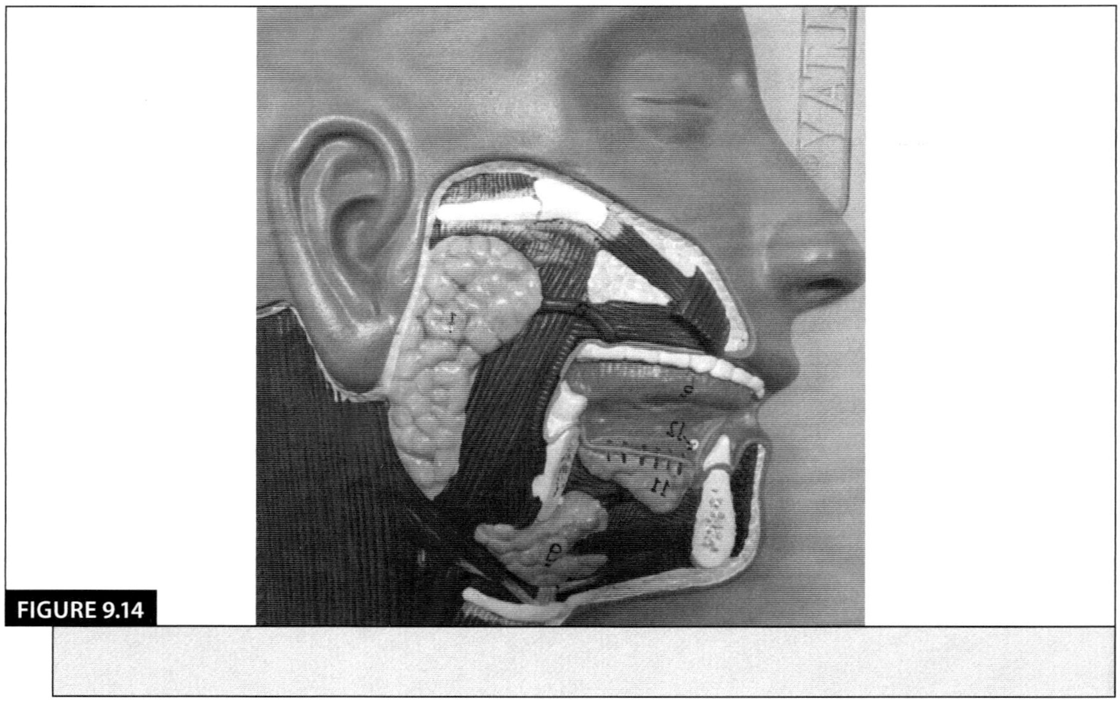

**FIGURE 9.14**

3. Label the three parts of the pharynx and the following structures on the diagrams below: esophagus, lower esophageal sphincter, stomach, trachea, and upper esophageal sphincter.

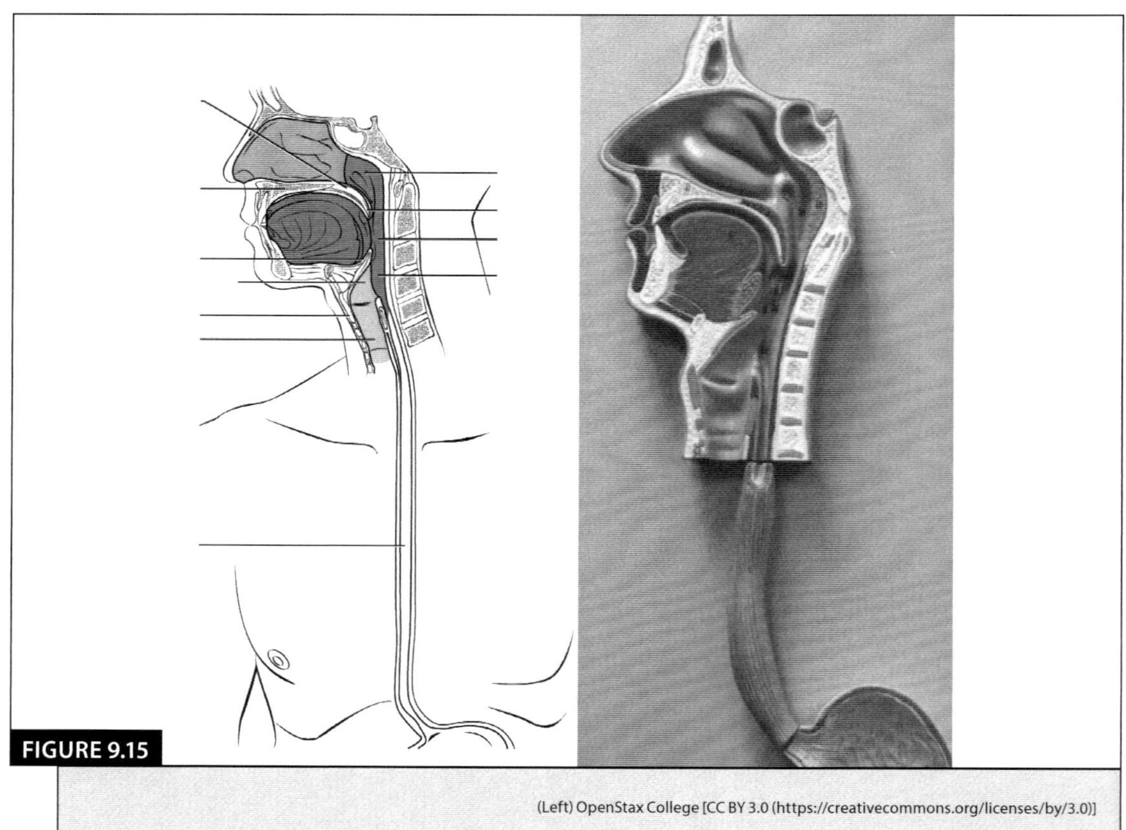

**FIGURE 9.15**

(Left) OpenStax College [CC BY 3.0 (https://creativecommons.org/licenses/by/3.0)]

**4.** Label the following parts of the stomach *using the model in lab and your text. Be sure to point to the structure and say it* **aloud** *then write it down.*

**body, cardia, circular muscle layer, esophagus, fundus, greater curvature, lesser curvature, longitudinal muscle layer, lower esophageal sphincter, oblique muscle layer, pyloric sphincter, pylorus,** and **rugae**

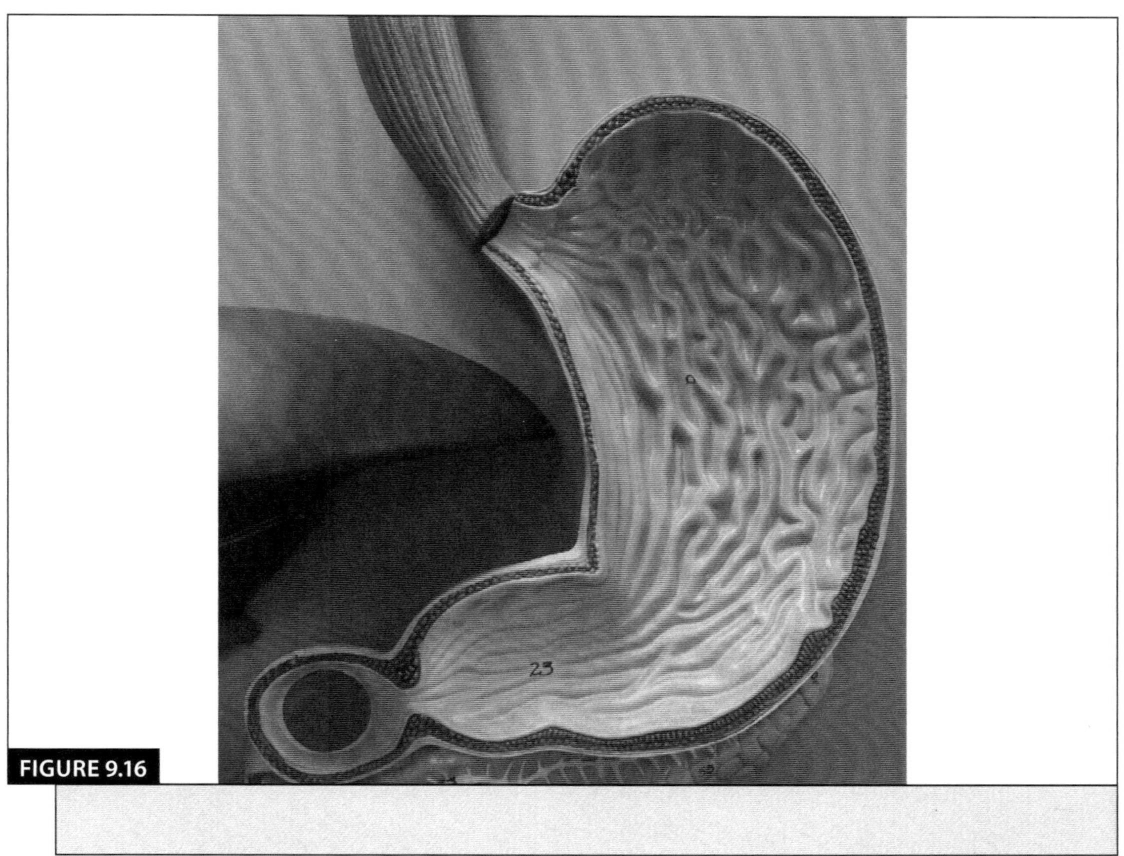

**FIGURE 9.16**

**5.** Label the following structures on the model below *using the model(s) in lab or dissection and text. When you write the term in place, say it aloud.*

**anal canal, anus, appendix, ascending colon, cecum, descending colon, duodenum, external anal sphincter, haustra, ileocecal valve, ileum, internal anal sphincter, jejunum, left colic flexure, plica circulare, rectum, right colic flexure, sigmoid colon, stomach,** and **transverse colon**

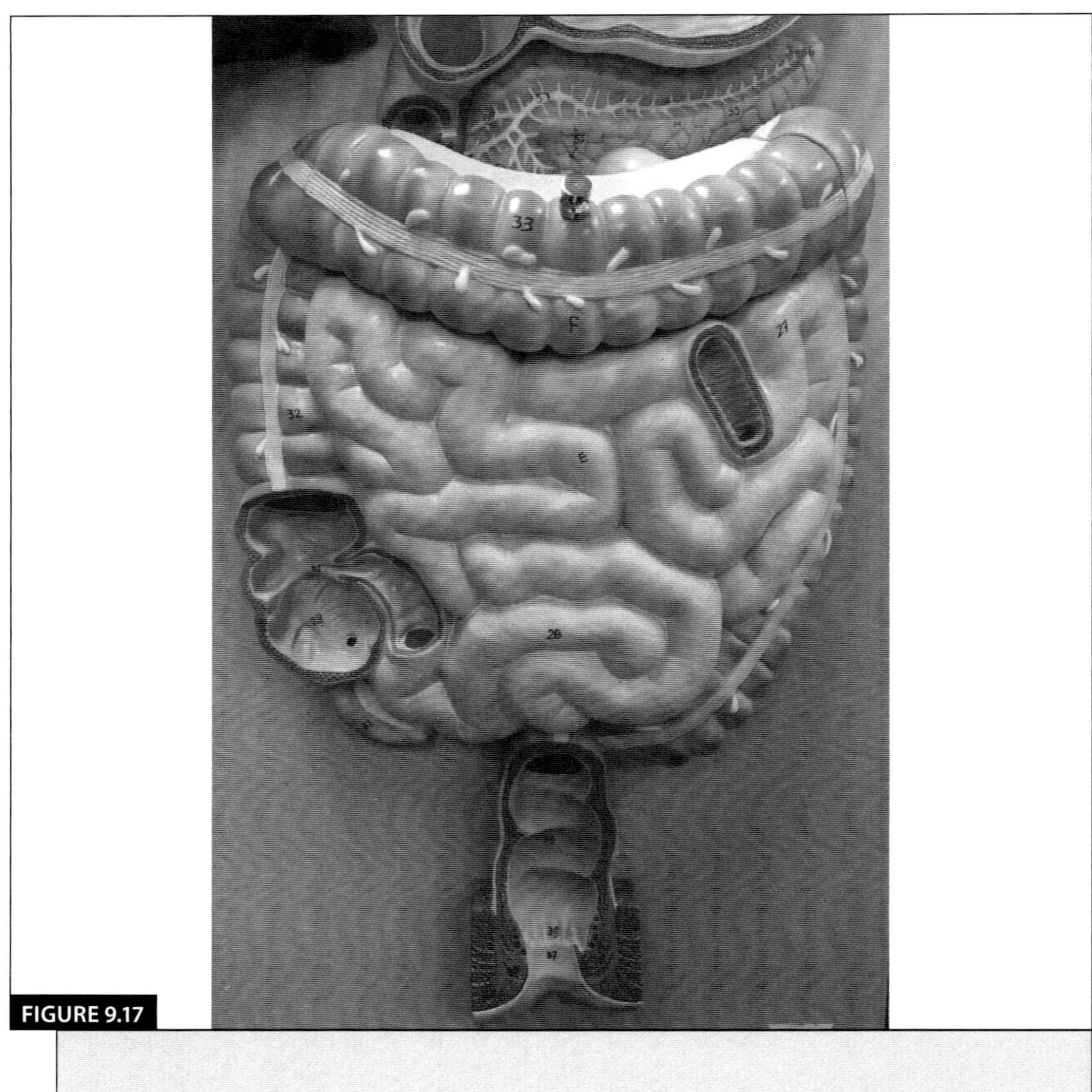

**FIGURE 9.17**

6. Label the figure of the descending colon below. **Teniae coli** (three ribbons of smooth muscle on the external colon), **epiploic appendages** (fatty deposits), and **haustra** (folds caused by the shorter teniae coli)

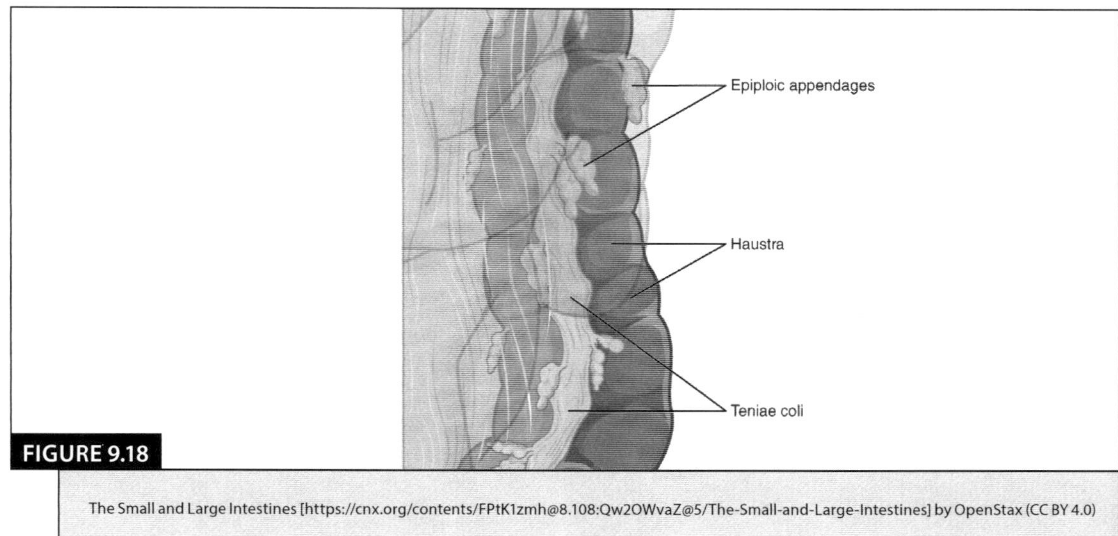


**FIGURE 9.18**

The Small and Large Intestines [https://cnx.org/contents/FPtK1zmh@8.108:Qw2OWvaZ@5/The-Small-and-Large-Intestines] by OpenStax (CC BY 4.0)

7. Label the structures on the model below: **accessory pancreatic duct, common bile duct, common hepatic duct, cystic duct, duodenum, duodenal papilla, gallbladder, hepatopancreatic ampulla, jejunum, left hepatic duct, primary pancreatic duct,** and **right hepatic duct**

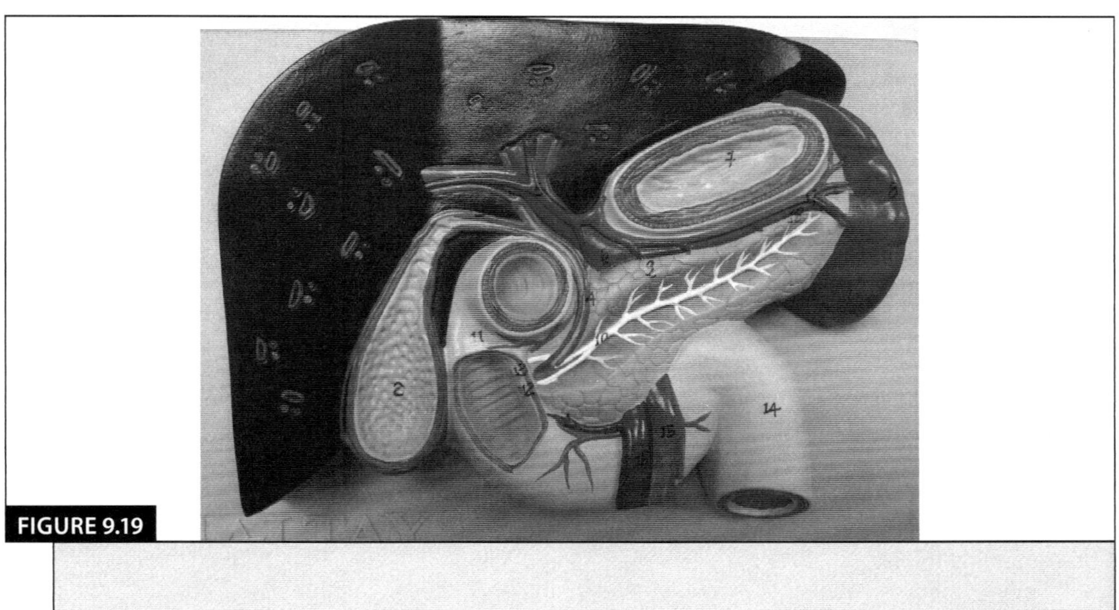

**FIGURE 9.19**

8. Label the structures on the figure below: **right hepatic duct, left hepatic duct, common hepatic duct, gallbladder, cystic duct, common bile duct, hepatopancreatic ampulla,** and **pancreas**

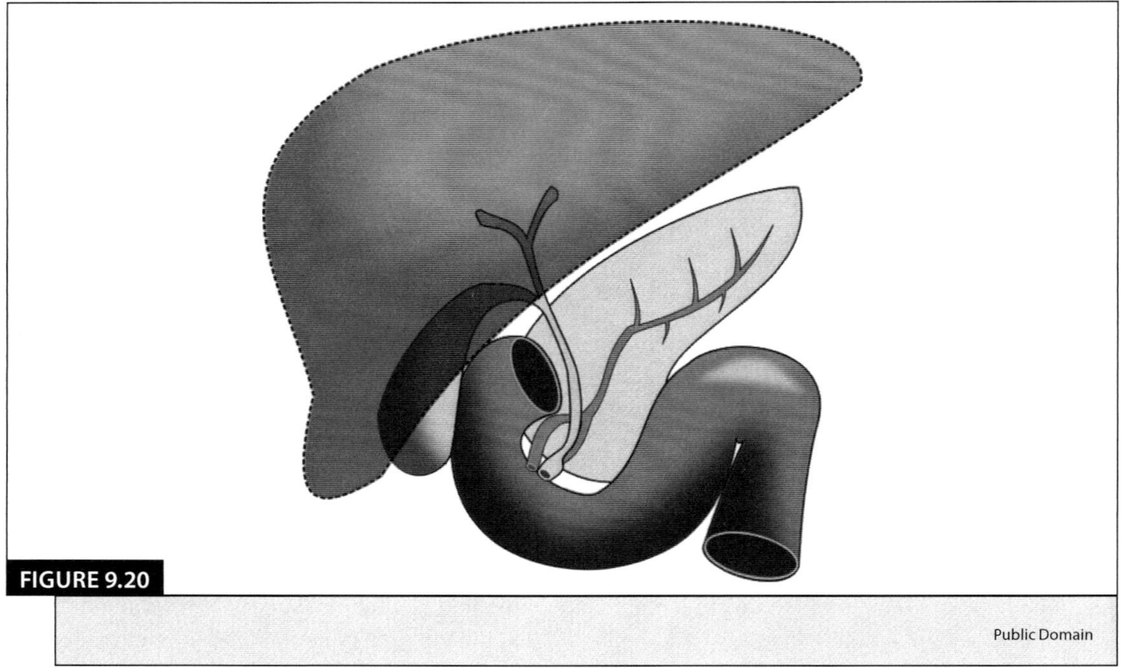

FIGURE 9.20

**9.** Obtain a compound light microscope and histology slides. Sketch what you see in the space below.

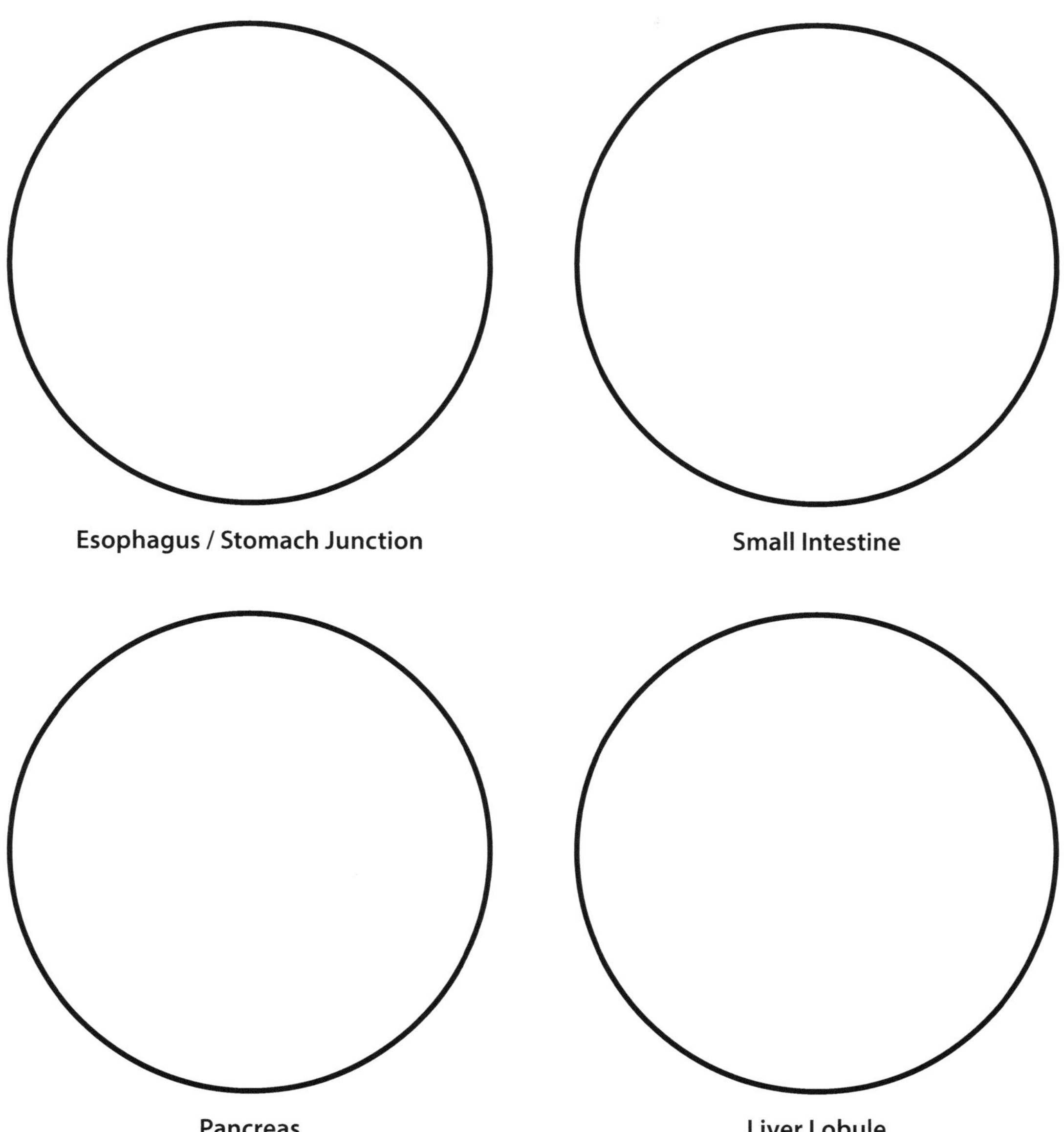

## EXTENSION QUESTIONS

*Complete these fully and in your own words.*

1. How would a hiatal hernia lead to heartburn and possible bleeding?

2. How does the Heimlich maneuver remedy choking if the obstruction is able to dislodge?

3. Answer all questions on the Carolina BioKit: Digestion handout and turn it in with this sheet.

*Remember to show these lab activities to the SSL instructor **before leaving** the lab. It will be marked off in the SSL book for grade credit towards your lab test.*

# 10

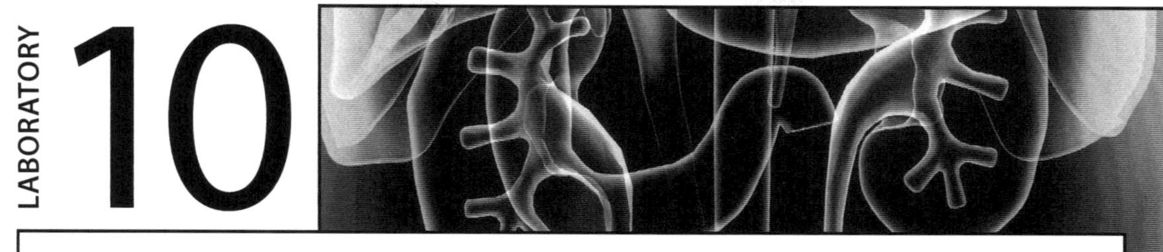

# URINARY SYSTEM
## PRE-LAB

Name: _________________________     Section: __________     Date: _________

## LEARNING OBJECTIVES

1. Identify and associate kidney structures with urinary system functions of waste removal, body fluid volumes, and body fluid composition.

2. Identify the blood vessels of the kidney and recognize their importance given the urinary system's functions.

3. Identify all parts of the duct system that deliver urine from the renal papilla to the external urethral orifice.

4. Learn to perform and interpret urinalysis.

*Checklist to complete* **before entering** *the science skills lab (SSL):*

☐ Actively read this packet of information.

☐ Complete the charts, tables or labeling and answer questions using your own words.

☐ Complete the electronic digital pre-lab quiz on Bb by Sunday.

☐ Review the attached anatomy list and take it to lab with you. Jot down key descriptive identifying words that aid in your lab test preparations.

## INTRODUCTION

The urinary system eliminates waste products by excretion. The elimination of waste helps the body maintain homeostasis. In the process of eliminating wastes, the urinary system manages homeostasis of body fluid volumes and fluid composition. Kidney functions are most dependent on the properties of osmosis and diffusion.

Osmosis and diffusion are major in regulating the concentrations of electrolytes in the body and maintaining normal pH of the blood. The major electrolytes in body fluids are ions comprised of four minerals that include sodium, ($Na^+$) potassium, ($K^+$), magnesium ($Mg^{2+}$), and calcium ($Ca^{2+}$). In addition to the four major electrolytes, blood carries proteins, enzymes, gas molecules, and metabolic wastes that determine the pH of the blood. Blood pH is mainly dictated by metabolism of cells and breathing.

Reasons for monitoring electrolytes are varied. Muscle cramping may involve improper hydration which leads to improper electrolyte replenishment. As we use our muscle, we need $Na^+$, $K^+$, and $Ca^{2+}$ to be present to start and stop action potentials. Without action potentials (electrical currents), the nervous system could not signal the muscle to contract. You could think of it like turning on the lights in your home. If there is no electricity, you will not have light. If there is no action potential, there is no muscle contraction and relaxation. The entire process is controlled by the minerals potassium, calcium, or magnesium in your diet. Without these minerals being dissolved in water and entering in and out of a person, muscles cramp due to mineral depletion.

Before the body reaches mineral depletion, some of the chemical by-products of muscle contraction become metabolic waste and enter the blood stream. One major by-product is carbon dioxide ($CO^2$). Along with $CO^2$ comes disassociated hydrogen ($H^+$) ion production. When they enter the blood stream, the blood becomes acidic. As $CO^2$ diffuses from the blood to the atmosphere in the respiratory system, carbonic anhydrase enzymes facilitate the conversion of H+ and bicarbonate to CO2 and water. This activity raises the pH of blood. In other words, the blood is made more alkaline (basic). Concurrent with the activities of the respiratory system and carbonic anhydrase, the urinary system is using the functional unit in the kidney (nephron) to remove more ($H^+$) protons by excretion, also making the blood more basic. These processes are important examples of homeostasis performed by the urinary system.

# ACTIVITY 1: REVIEW THE KIDNEY STRUCTURES IN FIGURE 10.1 BELOW AND LABEL THEM IN FIGURE 10.2

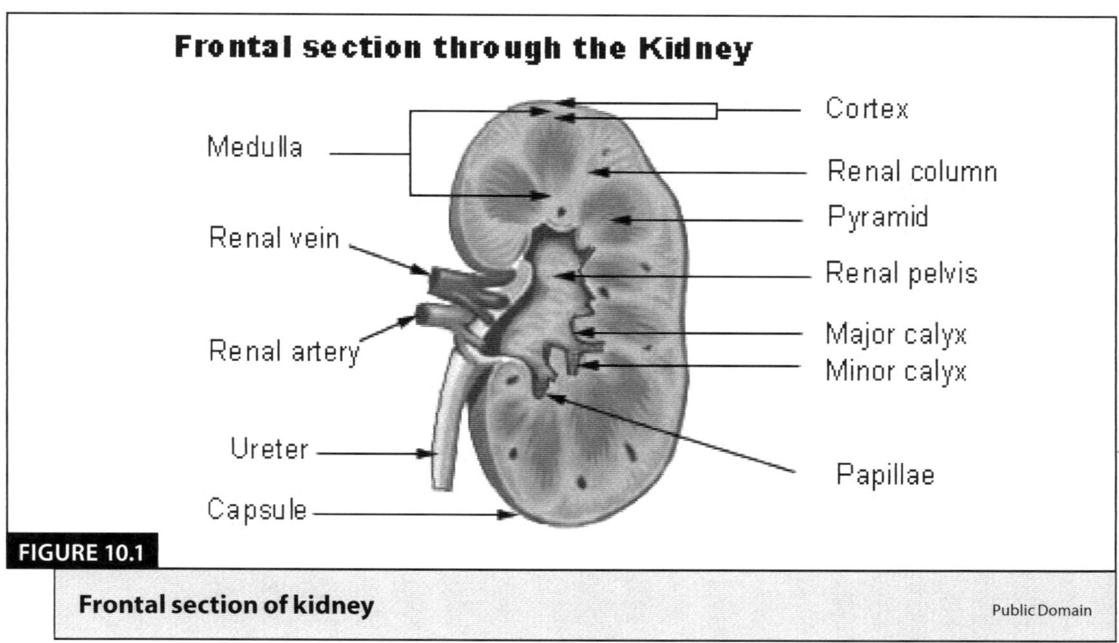


**FIGURE 10.1**

**Frontal section of kidney**



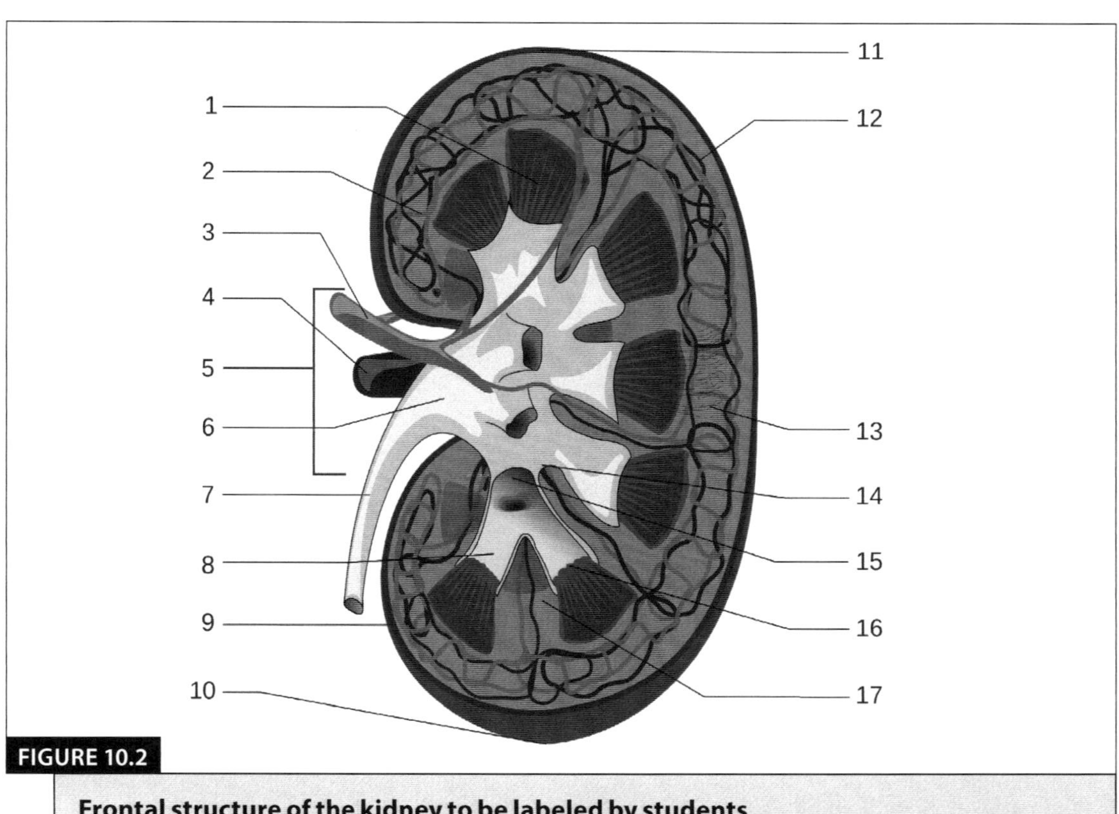


**FIGURE 10.2**

**Frontal structure of the kidney to be labeled by students**
Piotr Michał Jaworski; PioM EN DE PL [CC BY-SA 3.0 (http://creativecommons.org/licenses/by-sa/3.0/)]

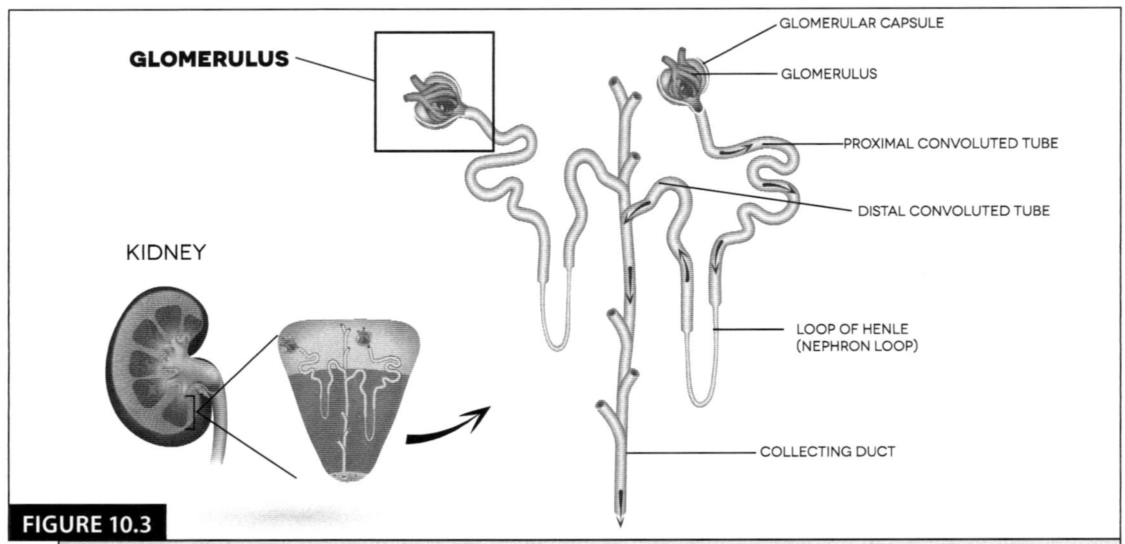


**FIGURE 10.3**

**Kidney pyramid showing two nephrons with a nephron zoomed in and labeled.**
Structures seen include the glomerulus, Bowman's capsule, PCT, Loop of Henle, DCT, collecting
duct, the cortex, and medulla.

https://en.wikipedia.org/wiki/Renal_pelvis

## ACTIVITY 2: REVIEW, DRAW, AND LABEL A JUXTAMEDULLARY NEPHRON AND ITS ASSOCIATED COLLECTING DUCT IN THE SPACE BELOW

The nephron and collecting duct are the principal urine forming structures. Draw in
the space below the upper/outer Cortex region, then the lower/inner Medulla region of
the kidney pyramids and respective cortical areas.

Cortex

Medulla

## ACTIVITY 3: USE THE ODIGIA TEXT OR ANY OPEN SOURCE RESOURCE TO COMPLETE THE FOLLOWING TABLE

**Anatomy and Physiology of the Kidney**

| TABLE 10.1 | | |
| --- | --- | --- |
| **STRUCTURE** | **FUNCTION** | **ADDITIONAL NOTES** |
| Glomerulus | | |
| Bowman Capsule | | |
| DCT | | |
| Descending Loop | | |
| Ascending Loop | | |
| PCT | | |
| Collection Tube | | |

## ACTIVITY 4: BEFORE YOU COME TO LAB, DEFINE THE FOLLOWING WORDS:

- Chronic kidney disease:

- Diabetes:

- Scurvy:

- Proteinuria:

- Amyloidosis:

*Remember to show this pre-lab to the SSL instructor* **before entering** *the lab. It will be marked off in the SSL book for grade credit towards your lab test.*

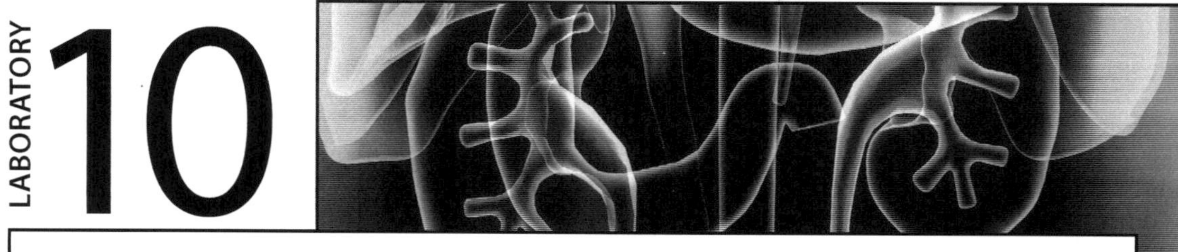

# 10

# URINARY SYSTEM
## IN-LAB ACTIVITIES

Name: _______________________  Section: __________  Date: _________

*In this lab you will be testing three simulated urine samples for appearance, odor, pH, sugar, and proteins. Three patients' case studies will be used to help you understand the basics of urinalysis as a tool to aid in diagnosis for medical conditions, and one normal urine sample will be tested to use as a control.*

## LAB ACTIVITY

1. Analyze the four simulated urine samples and fill in the corresponding tables with your findings.

2. Compare your observations to the charts, detailing possible disorders according to normal values.

3. Based on these results, as well as the case studies, see if you can provide a possible diagnosis of the patient's condition.

## LAB QUESTIONS

*As you perform this lab, you will be asked to find the answer to the following questions:*

1. **Analyze** why it is important to do a urinalysis as a diagnostic test for medical diseases.

2. Tentatively give to each one of the patient's urine samples, the possible **diagnostic** for their medical condition based on your observations. Some patients' conditions may include diabetes, proteinuria, kidney disease, infection, dehydration, diabetic ketoacidosis, etc.

## PATIENT 1 (URINE SAMPLE #1)

Jeff Jones is 19 years old. He notices that he has increased urine output (polyuria), increased appetite (polyphagia), and great thirst (polydipsia). He has also experienced unexplained weight loss.

## PATIENT 2 (URINE SAMPLE #2)

Mr. Thomson is 60 years old and has been unusually tired for several weeks. He occasionally feels dizzy and lately he finds it increasingly difficult to sleep at night. He has swollen ankles and feet and his face looks puffy. He experiences a burning pain in his lower back, just below the rib cage. He also notices that his urine is dark in color. He goes to see his physician, who finds that he has elevated blood pressure, and that the kidney region is sensitive to pressure.

## PATIENT 3 (URINE SAMPLE #3)

Ms. Smith is 27 years old and has been experiencing painful and difficult urination (dysuria), frequency of urination and urgency. Her urine has a milky color. She also has fever and malaise, which is evidence of infection. Upon seeking treatment, she is given antibiotic therapy. After a few days on antibiotics, her symptoms disappear.

## PATIENT 4—NORMAL URINE SAMPLE #4 (CONTROL)

# ACTIVITY 1: PHYSICAL CHARACTERISTIC OBSERVATIONS ——

*Perform the following test on each urine sample. See the available urinalysis kit in the lab.*

1. **Gross visual inspection** (Describe the *color* and the *clarity* of each urine sample using the terms from your Pre-lab.(

2. **Odor or smell** (*odorless, mild, strong, fruity, sweet, vinegar, ammonia, fishy, fetid (stinky)*)

3. **pH—Chemical examination/urine dipstick** (*pH values for each sample*)

## MATERIALS YOU WILL NEED

- 4 urine samples (patient 1 to patient 4)
- 8 glass test tubes (16x100mm) (or small cups if in the lab)
- 2 test tube racks
- 4 test tube holders
- 4 graduated 1 ml. dropper plastic pipets
- 4 pH test trips and 1 pH chart for analysis (or a pH meter if in the lab)
- A marker, crayon, or tape for labeling each tube
- Exam globes of your hand's size (S, M, L)
- A disposable bag/container
- Paper towels/surface cleaner to clean your work area at the end

## STEPS FOR OBSERVING PHYSICAL CHARACTERISTICS AND pH:

1. Have your lab coat on and wear your protecting gloves.

2. Label 4 lab tubes 1G, 2G, 3G, and 4G, (G for Glucose) and place them standing on the test tube rack. These will be used for Activity 1 and Activity 2.

3. Label 4 lab tubes 1P, 2P, 3P, and 4P, (P for Proteins) and place them in another row. To be used later for Activity 3.

4. Carefully shake each patient urine sample thoroughly as you pipet and dispense 3 mL (3 cc) of each sample into the properly labeled series "G" test tubes. Use a different pipet for each urine sample. Do not mix the pipets; you might want to label them as well. Keep them standing in order on a test tube close to each patient's urine sample. You will be needing them again for Activity 2. Dispense Patient 1 urine sample on test tube 1G, dispense Patient 2 urine sample on test tube 2G, Patient 3 urine sample on test tube 3G, and Patient 4 urine sample on test tube 4G.

5. For each urine sample provided, observe and record the color, clarity, and smell of the urine by using the terms you studied above.

6. Record your findings in Table 10.2 results and analysis for each tube.

7. Proceed to dip a pH test strip into the test tube 1G, carefully tilting the tube but not to the point of causing a spill. Immediately compare the color changes of the test strip to the comparator chart within 30 seconds of sampling. Record the pH value in Table 10.2.

8. Repeat Step 6 for the other test tubes and record your findings on the pH column.

**Physical characteristics and pH results and analysis**

| TABLE 10.2 | | | | |
|---|---|---|---|---|
| SAMPLE | COLOR | CLARITY | SMELL | PH |
| Patient 1, Tube 1G | | | | |
| Patient 2, Tube 2G | | | | |
| Patient 3, Tube 3G | | | | |
| Patient 4, Tube 4G | | | | |

## ACTIVITY 2: TESTING FOR SUGAR (BENEDICT'S TEST)

Benedict's solution, when mixed with urine, modifies the color of the urine. This is **heated** so must be performed with glass tubes or containers.

### MATERIALS FOR TESTING FOR THE PRESENCE OF SUGARS:

- The same 4 1G to 4G urine tubes with 3 mL urine samples used previously
- The same 4 pipets used previously– do not mix them up.
- The Benedict's solution
- 1 new pipet to dispense the Benedict's solution into each test tube
- 4 test tube holders
- Another test tube rack
- A pair of heat protecting gloves
- A 400 ml beaker
- A hot plate equipment
- 200 ml of tap water
- A timer device

### STEPS FOR ACTIVITY 2: TESTING FOR SUGARS

1. Place 200 mL of water into a 400 mL beaker and place the beaker on the hot plate to create a hot water bath. Set it on, and allow it to stay on a mild boiling state. (Be mindful of the hot water. Consider your time since it might take 30 minutes or so to bring the water to quick boiling point. To save time, you could use hot water from the sink to start with.)

2. Using the new clean pipet, into each 1G to 4G tubes, pipette approximately 2 mL of Benedict's solution. Do not touch the urine samples with the tip of the pipet.

3. Record the color of the solution in Table 2 on column "color before heating" for each 1G to 4G tubes.

4. Using the tube holders, proceed to carefully place the tubes 1G to 4 G into the hot water bath and allow it to boil for 2 minutes using your timer device. After the 2 minutes, turn the hot plate off, and while wearing your heat resistant gloves, remove the hot bath from the hot plate. Be very careful.

5. Using the test tube holders, observe each test tube for any color changes for patients 1 to 4, and record your observations in the column "color after heating 2 minutes."

6. A positive reaction will result in a yellow to red color. Note if it was a positive or negative reaction.

**Benedict's test for the presence of sugar**

| TABLE 10.3 | | | |
|---|---|---|---|
| SAMPLE | COLOR BEFORE HEATING | COLOR AFTER HEATING 2 MINUTES | RESULT (+ OR −) |
| Patient 1, Tube 1G | | | |
| Patient 2, Tube 2G | | | |
| Patient 3, Tube 3G | | | |
| Patient 4, Tube 4G | | | |

## ACTIVITY 3: TESTING FOR THE PRESENCE OF PROTEINS—BIURET TEST

### MATERIALS NEEDED FOR TESTING FOR PROTEINS

- The 4 glass test tubes labeled 1P, 2P, 3P and 4 P with the 3 mL urine samples each for patient 1 to 4.
- The 4 pipets that were used to dispense the urine samples into each tube
- The Biuret solution
- 1 clean, new 1 mL pipet

## STEPS FOR ACTIVITY 3: TESTING FOR PROTEINS

1. Carefully shake each patient urine sample thoroughly as you pipet and dispense 3 mL (3 cc) of each sample into the properly labeled series "P" test tubes. Use a different pipet for each urine sample. Do not mix the pipets; you might want to label them as well. Keep them standing in order on a test tube close to each patient's urine sample. Dispense Patient 1 Urine sample on test tube 1P, dispense Patient 2 urine sample on test tube 2P, Patient 3 urine sample on test tube 3P, and Patient 4 urine sample on test tube 4P. Place them in order on the test tube rack.

2. Observe the initial color sample for each tube and record it on Table 10.4 **"initial color of sample."**

3. Add 2 mL (2 cc) of Biuret solution to each patient's tube. Do not touch the urine sample with this pipet dispensing the Biuret solution.

4. Gently swirl each tube to mix the urine samples with the Biuret solution.

5. Observe the final color after adding Biuret solution to each tube and record it on Table 3 **final color of sample. Note your RESULTS if + or –.**

6. After finishing your observations, rinse and put away all equipment. Dispose appropriately of all glassware to air dry, pipets, and urine samples following standard laboratory procedures.

7. Clean your area, and orderly set aside the urinary samples for other classmates to use in the designated area.

**Biuret test for the presence of proteins**

| TABLE 10.4 | | | |
|---|---|---|---|
| SAMPLE | INITIAL COLOR OF SAMPLE | FINAL COLOR OF SAMPLE WITH BIURET SOLUTION | RESULT (+ OR–) |
| Patient 1, Tube 1P | | | |
| Patient 2, Tube 2P | | | |
| Patient 3, Tube 3P | | | |
| Patient 4, Tube 4P | | | |

## ACTIVITY 4: ANSWER THE QUESTIONS PRESENTED TO YOU AT THE START OF THIS LAB

*Use your Pre-lab to find the information.*

1. Analyze why it is important to do a urinalysis as a diagnostic test for medical diseases.

2. List some medical disorders as possible diagnosis for Patient 1 urine sample.

3. List some medical disorders as possible diagnosis for Patient 2 urine sample.

4. List some medical disorders as possible diagnosis for Patient 3 urine sample.

5. List some medical disorders as possible diagnosis for Patient 4 urine sample.

6. Why do you use safety gloves and perform these tests carefully (ex: Do not touch the urine samples with the tip of the pipet.)?

7. Write the sequence correctly that traces the path of urine after it leaves the kidneys in males and females.

8. How did the pH change between samples? What does the change in pH indicate?

# 11

# MALE REPRODUCTIVE SYSTEM
## PRE-LAB

Name: ______________________     Section: __________     Date: _________

*Print this Lab handout to complete before attending your lab time, or fill in to return to your instructor or Lab TA prior to receiving the Lab handout.*

## LEARNING OBJECTIVES

1. Identify, locate, and describe the function of the structures in the male reproductive systems.

2. Examine the processes of spermatogenesis.

3. Discuss the hormonal control of the male reproductive system.

*Checklist to complete* **before entering** *the science skills lab (SSL):*

☐ Actively read this packet of information.

☐ Complete the charts, tables or labeling and answer questions using your own words.

☐ Complete the electronic digital pre-lab quiz on Bb by Sunday.

☐ Review the attached anatomy list and take it to lab with you. Jot down key descriptive identifying words that aid in your lab test preparations.

## ACTIVITY 1: IDENTIFY, LOCATE, AND DESCRIBE THE FUNCTION OF THE STRUCTURES IN THE MALE REPRODUCTIVE SYSTEMS

*Read the content in your text and the introductory paragraphs below. Then, complete Activity 1 in lab.*

The function of the male reproductive system is to produce sperm and transfer them to the female reproductive tract. The paired testes are crucial components in this process, as they produce both sperm and androgens (hormones that support male reproductive physiology). In humans, the most important male androgen is testosterone. Several accessory organs and ducts aid the process of sperm maturation and transport the sperm and other seminal components to the penis, which delivers sperm to the female reproductive tract.

**Spermatogenesis,** the production of sperm, occurs within the **seminiferous tubules** that make up most of the **testis.** The scrotum is the muscular sac that holds the testes outside of the body cavity.

Spermatogenesis begins with mitotic division of spermatogonia (stem cells) to produce primary spermatocytes that undergo the two divisions of meiosis to become second-ary spermatocytes, then the haploid spermatids. During spermiogenesis, the haploid spermatids are transformed ("mature") into spermatozoa (formed sperm). Upon release from the seminiferous tubules, sperm are moved to the epididymis where they continue to mature. During ejaculation, the sperm next exit the **epididymis** through the **ductus deferens,** a duct in the spermatic cord that leaves the **scrotum.** The ductus deferens meets the **seminal vesicle,** a gland that contributes fructose and proteins, at the **ejacu-latory duct.** The fluid continues through the **prostatic urethra,** where secretions from the **prostate** are added to form semen. These secretions help the sperm to travel through the urethra and into the female reproductive tract. Secretions from the **bulbourethral glands** protect sperm by cleansing and lubricating the penile (spongy) urethra.

The **penis** is the male organ of copulation. Columns of erectile tissue called the **corpora cavernosa** and **corpus spongiosum** (surrounds the spongy urethra) fill with blood when sexual arousal activates vasodilatation in the blood vessels of the penis.

From your reading the descriptions here, **label** the figure below.

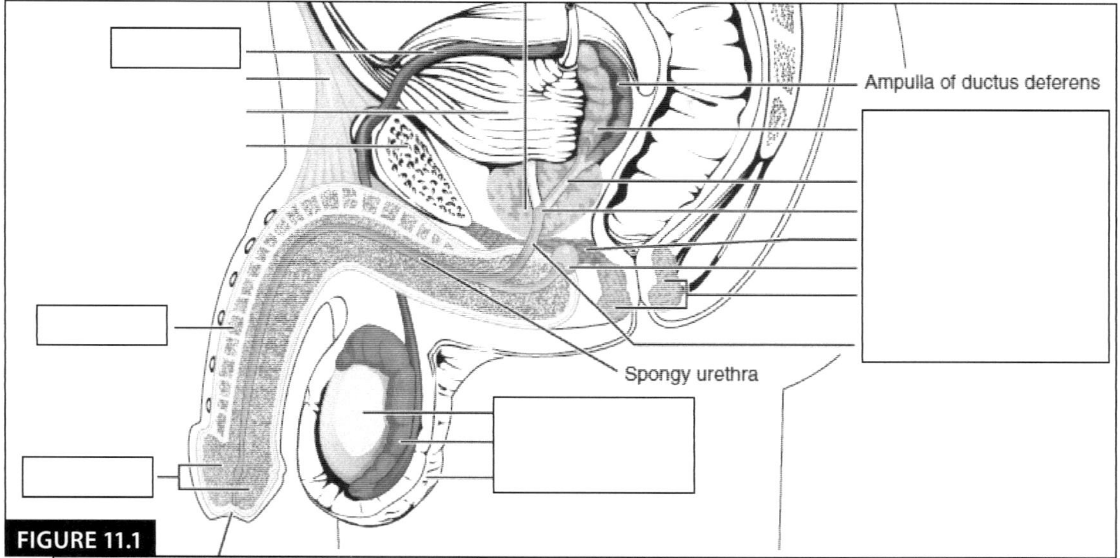



**FIGURE 11.1**

**The structures of the male reproductive system include the *testes*, the *epididymides*, the *penis*, and the ducts and glands that produce and carry semen.** Sperm exit the scrotum through the *ductus deferens,* which is bundled in the spermatic cord. The *seminal vesicles* and *prostate gland* add fluids to the sperm to create semen. Anatomy and Physiology of the Male Reproductive System [https://cnx.org/contents/FPtK1zmh@8.108:Nw1tEY4R@9/Anatomy-and-Physiology-of-the-Male-Reproductive-System] by OpenStax, (CC BY 4.0)

## SCROTUM

The testes are located in a skin-covered, highly pigmented, muscular sack called the **scrotum** that extends from the body behind the penis. This location is important in sperm production, which occurs within the testes, and proceeds more efficiently when the testes are kept 2 to 4°C below core body temperature. The **dartos muscle** inserts on the scrotal skin causing the scrotum to wrinkle or relax, which alters surface area and heat loss. Descending from the internal oblique muscle of the abdominal wall are the two cremaster muscles, which cover each testis like a muscular net. The **cremaster muscles** can elevate the testes in cold weather (or water), moving the testes closer to the body to retain heat.

## TESTES

The **testes** (singular = testis) are the male **gonads**—that is, the male reproductive organs. They produce both sperm and androgens, such as testosterone, and are active throughout the reproductive lifespan of the male. The outer covering of the testes is the tunica vaginalis and beneath the tunica vaginalis is the tunica albuginea—a tough, white, dense connective tissue layer covering the testis itself. Not only does the tunica albuginea cover the outside of the testis, it also invaginates to form septa that divide the testis into 300 to 400 structures called lobules. Within the lobules, sperm develop in structures called seminiferous tubules.

The tightly coiled **seminiferous tubules** form the bulk of each testis. They are composed of developing sperm cells surrounding a lumen, the hollow center of the tubule, where formed sperm are released into the duct system of the testis. Specifically, from the lumen of the seminiferous tubules, sperm move into the straight tubules and from there, into a fine meshwork of tubules called the rete testes. Sperm leave the rete testes, and the testis itself, through the 15 to 20 efferent ductules that cross the tunica albuginea into the epididymis.

**Label** the figure below.

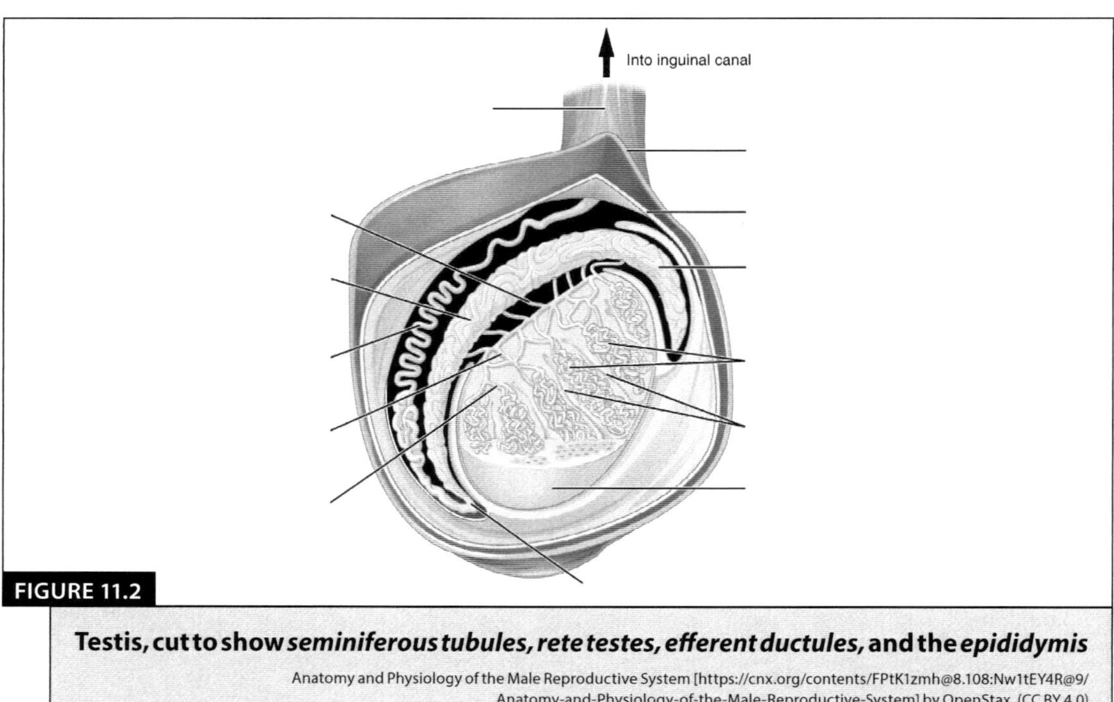


**FIGURE 11.2**

**Testis, cut to show *seminiferous tubules, rete testes, efferent ductules,* and the *epididymis***

Anatomy and Physiology of the Male Reproductive System [https://cnx.org/contents/FPtK1zmh@8.108:Nw1tEY4R@9/
Anatomy-and-Physiology-of-the-Male-Reproductive-System] by OpenStax, (CC BY 4.0)

## EPIDIDYMIS

From the lumen of the seminiferous tubules, the immotile sperm are surrounded by testicular fluid and moved to the **epididymis,** a coiled tube attached to the testis. As they are moved along the length of the epididymis, the sperm further mature and acquire the ability to move under their own power. Once inside the female reproductive tract, they will use this ability to move independently toward the unfertilized egg. The more mature sperm are then stored in the tail of the epididymis (the final section) until ejaculation occurs.

## DUCT SYSTEM

During ejaculation, sperm exit the tail of the epididymis and are pushed by smooth muscle contraction to the **ductus deferens** (also called the vas deferens). The ductus deferens is a thick, muscular tube that is bundled together inside the scrotum with connective tissue, blood vessels, and nerves into a structure called the **spermatic cord.** From each epididymis, each ductus deferens extends superiorly into the abdominal cavity through the **inguinal canal** in the abdominal wall. From here, the ductus deferens continues posteriorly to the pelvic cavity, ending posterior to the bladder where it dilates in a region called the ampulla. Sperm make up only 5 percent of the final volume of **semen,** the thick, milky fluid that the male ejaculates. The bulk of semen is produced by three critical accessory glands of the male reproductive system: the seminal vesicles, the prostate, and the bulbourethral glands.

## SEMINAL VESICLES

As sperm pass through the ampulla of the ductus deferens at ejaculation, they mix with fluid from the associated **seminal vesicle.** The paired seminal vesicles are glands that contribute approximately 60 percent of the semen volume. Seminal vesicle fluid contains large amounts of fructose, which is used by the sperm mitochondria to generate ATP to allow movement through the female reproductive tract. The fluid, now containing both sperm and seminal vesicle secretions, next moves into the associated **ejaculatory duct,** a short structure formed from the ampulla of the ductus deferens and the duct of the seminal vesicle. The paired ejaculatory ducts transport the seminal fluid into the next structure, the prostate gland.

## PROSTATE GLAND

The centrally located **prostate gland** sits anterior to the rectum at the base of the bladder surrounding the prostatic urethra (the portion of the urethra that runs within the prostate). About the size of a walnut, the prostate excretes an alkaline, milky fluid to the passing seminal fluid—now called semen—that is critical to first coagulate and then decoagulate the semen following ejaculation. The temporary thickening of semen helps retain it within the female reproductive tract, providing time for sperm to utilize the fructose provided by seminal vesicle secretions. When the semen regains its fluid state, sperm can then pass farther into the female reproductive tract.

## BULBOURETHRAL GLANDS

The final addition to semen is made by two **bulbourethral glands** (or Cowper's glands) that release a thick, salty fluid that lubricates the end of the urethra and the vagina, and helps to clean urine residues from the penile urethra.

## THE PENIS

The **penis** is the male organ of copulation (sexual intercourse). The shaft of the penis surrounds the urethra and is composed of three column-like chambers of erectile tissue that span the length of the shaft. Each of the two larger lateral chambers is called a **corpus cavernosum** (plural = corpora cavernosa). Together, these make up the bulk of the penis. The **corpus spongiosum,** which can be felt as a raised ridge on the erect penis, is a smaller chamber that surrounds the spongy, or penile, urethra. The end of the penis, called the **glans penis,** has a high concentration of nerve endings, resulting in very sensitive skin that influences the likelihood of ejaculation. The skin from the shaft extends down over the glans and forms a collar called the **prepuce** (or foreskin).

**Label** the figure below.

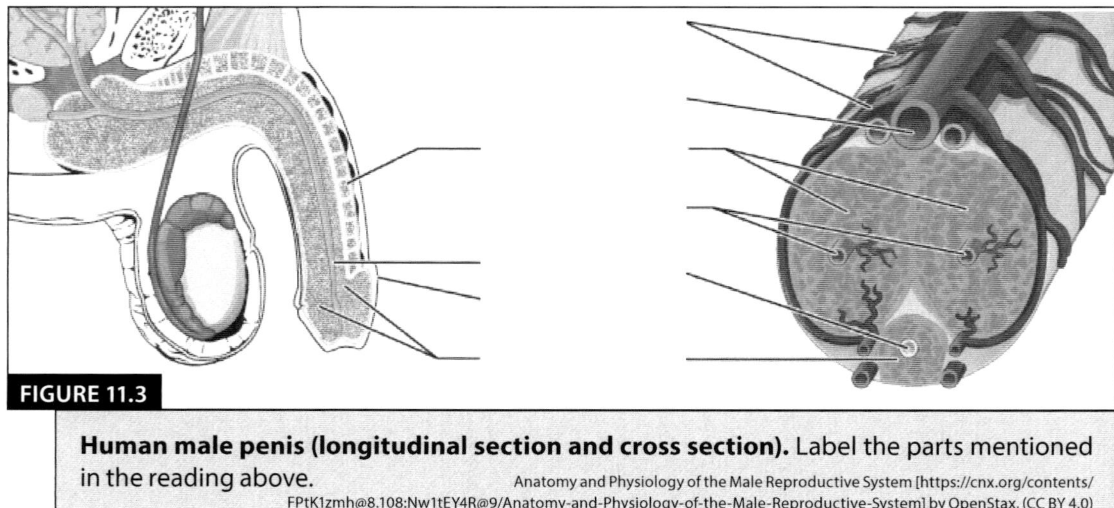



**Human male penis (longitudinal section and cross section).** Label the parts mentioned in the reading above. Anatomy and Physiology of the Male Reproductive System [https://cnx.org/contents/FPtK1zmh@8.108:Nw1tEY4R@9/Anatomy-and-Physiology-of-the-Male-Reproductive-System] by OpenStax, (CC BY 4.0)

# ACTIVITY 2: EXAMINE THE PROCESSES OF SPERMATOGENESIS

*Read the content in your text and the introductory paragraphs below. Then, complete Activity 2 in lab.*

As just noted, spermatogenesis occurs in the seminiferous tubules that form the bulk of each testis. The process begins at puberty, after which time sperm are produced constantly throughout a man's life. One production cycle, from spermatogonia through formed sperm, takes approximately 64 days. The process of spermatogenesis begins with mitosis of the diploid spermatogonia. Because these cells are diploid ($2n$), they each have a complete copy of the father's genetic material, or 46 chromosomes. However, mature gametes are haploid ($1n$), containing 23 chromosomes—meaning that daughter cells of spermatogonia must undergo a second cellular division through the process of meiosis.

The process of spermatogenesis is as follows. Two identical diploid cells result from spermatogonia mitosis. One of these cells remains a spermatogonium (stem cell), and the other becomes a primary **spermatocyte,** the next stage in the process of spermatogenesis. As in mitosis, DNA is replicated in a primary spermatocyte, and the cell undergoes cell division to produce two cells with identical chromosomes. Each of these is a secondary spermatocyte. Now a second round of cell division occurs in both of the secondary spermatocytes, separating the chromosome pairs. This second meiotic division results in a total of four cells with only half of the number of chromosomes. Each of these new cells is a **spermatid.** Although haploid, early spermatids look very similar to cells in the earlier stages of spermatogenesis, with a round shape, central nucleus, and large amount of cytoplasm. A process called **spermiogenesis** transforms these early spermatids, reducing the cytoplasm, and beginning the formation of the parts of a true sperm. The fifth stage of germ cell formation—spermatozoa, or formed sperm—is the end result of this process, which occurs in the portion of the tubule nearest the lumen. Eventually, the sperm are released into the lumen and are moved along a series of ducts in the testis toward a structure called the epididymis for the next step of sperm maturation.

Write in the cells and processes correctly in the flow chart to the right.

1. Spermatogonium

2. Spermatozoa

3. Primary spermatocyte

4. Secondary spermatocyte

5. Spermatid

6. Mitosis

7. Meiosis I

8. Meiosis II

9. Spermiogenesis

## STRUCTURE OF FORMED SPERM

Sperm are smaller than most cells in the body; in fact, the volume of a sperm cell is 85,000 times less than that of the female gamete. Sperm have a distinctive **head, mid-piece,** and **tail** region. The head of the sperm contains the extremely compact **haploid nucleus** (DNA) with very little cytoplasm. A structure called the **acrosome** covers most of the head of

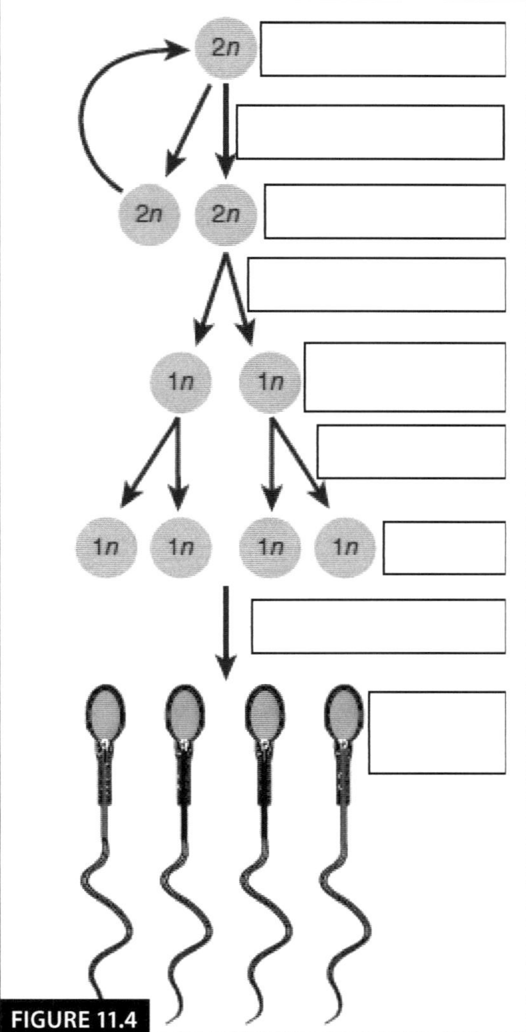

**FIGURE 11.4**

**Process of Meiosis through sperm maturation.** Match the correct terms with the steps in the process from your reading above. Anatomy and Physiology of the Male Reproductive System [https://cnx.org/contents/FPtK1zmh@8.108:Nw1tEY4R@9/Anatomy-and-Physiology-of-the-Male-Reproductive-System] by OpenStax, (CC BY 4.0)

the sperm cell as a "cap" that is filled with lysosomal enzymes important for preparing sperm to participate in fertilization. Tightly packed **mitochondria** fill the mid-piece of the sperm. ATP produced by these mitochondria will power the **flagellum,** which extends from the neck and the mid-piece through the tail of the sperm, enabling it to move the entire sperm cell.

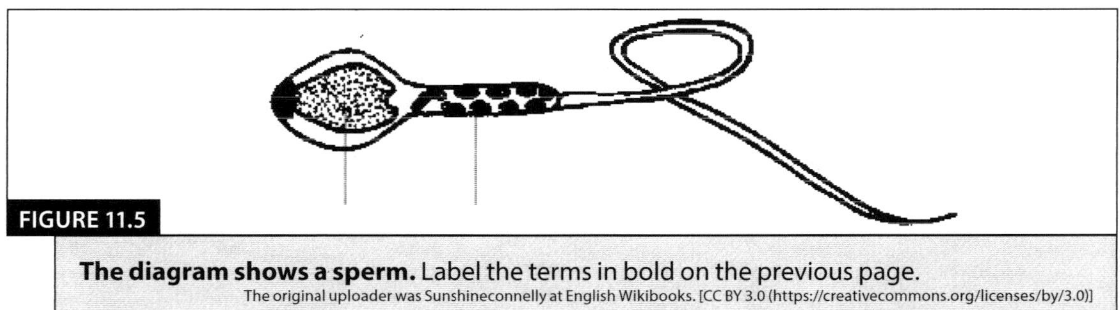



**The diagram shows a sperm.** Label the terms in bold on the previous page.
The original uploader was Sunshineconnelly at English Wikibooks. [CC BY 3.0 (https://creativecommons.org/licenses/by/3.0)]

## ACTIVITY 3: DISCUSS THE HORMONAL CONTROL OF THE MALE REPRODUCTIVE SYSTEM

*Read the content in your text and the introductory paragraphs below. Then, complete Activity 3 in lab.*

### TESTOSTERONE

Testosterone, an androgen, is a steroid hormone produced by **Leydig cells.** The alternate term for Leydig cells, interstitial cells, reflects their location between the seminiferous tubules in the testes. In male embryos, testosterone is secreted by Leydig cells by the seventh week of development, with peak concentrations reached in the second trimester. This early release of testosterone results in the anatomical differentiation of the male sexual organs. In childhood, testosterone concentrations are low. They increase during puberty, activating characteristic physical changes and initiating spermatogenesis.

1.  Describe the control of testosterone release below. Include in your answer the following structures and hormones: hypothalamus, adenohypophysis, testis, GnRH, FSH, LH, testosterone, and inhibin.

**2.** Is this process negative or positive feedback?

_______________________________________________

_______________________________________________

*Checklist to complete* **before entering** *the science skills lab (SSL):*

☐ Actively read this packet of information.

☐ Complete the charts, tables or labeling and answer questions using your own words.

☐ Complete the electronic digital pre-lab quiz on Bb by Sunday.

☐ Review the attached anatomy list and take it to lab with you. Jot down key descriptive identifying words that aid in your lab test preparations.

## ANATOMY LIST

**1.** Identify the following structures on the male models:

   **a.** scrotum

      **i.** cremaster muscle

   **b.** testis (plural—testes)

      **i.** tunica albuginea

      **ii.** seminiferous tubules

   **c.** rete testis

   **d.** epididymis

      **i.** head

      **ii.** body

      **iii.** tail

   **e.** ductus deferens (vas deferens)

   **f.** spermatic cord

      **i.** pampiniform plexus

   **g.** seminal vesicle

   **h.** ejaculatory duct

   **i.** prostate gland

   **j.** bulbourethral gland (Cowper's gland)

    **k.** penis

        **i.** corpus spongiosum

        **ii.** corpus cavernosum

        **iii.** prepuce of the penis (foreskin)

        **iv.** glans penis

**2.** Identify the following microscopic structures on a slide of the testis:

    **a.** interstitial cells (cells of Leydig)

    **b.** seminiferous tubules

        **i.** spermatozoa

*Remember to show this pre-lab to the SSL instructor* **before entering** *the lab. It will be marked off in the SSL book for grade credit towards your lab test.*

# 11
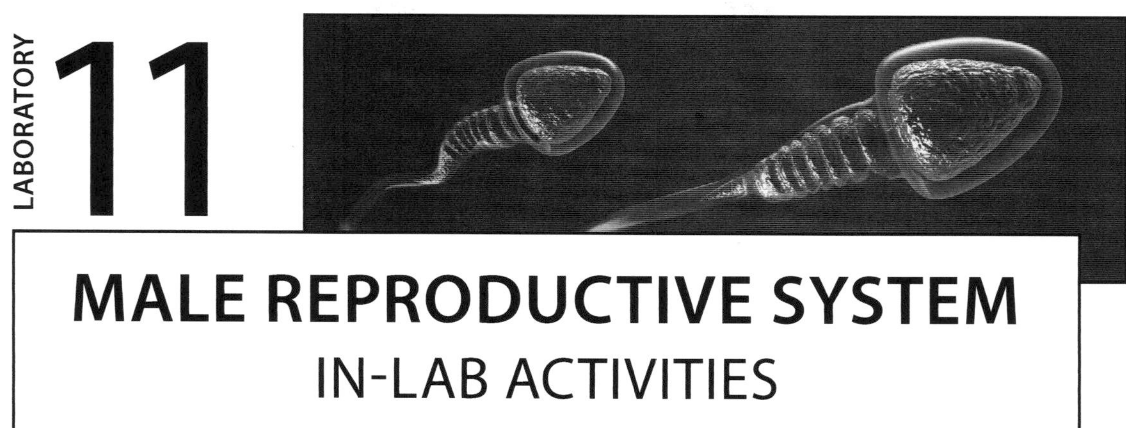

# MALE REPRODUCTIVE SYSTEM
## IN-LAB ACTIVITIES

Name: _______________________    Section: ___________    Date: _________

## LEARNING OBJECTIVES

1. Identify, locate, and describe the function of the structures in the male reproductive systems.

2. Examine the processes of spermatogenesis.

3. Discuss the hormonal control of the male reproductive system.

## ACTIVITY 1: IDENTIFY, LOCATE, AND DESCRIBE THE FUNCTION OF THE STRUCTURES IN THE MALE REPRODUCTIVE SYSTEMS

*You will need a male reproductive model (if available) to complete this exercise.*

1.  Label the missing structures of the figure below.

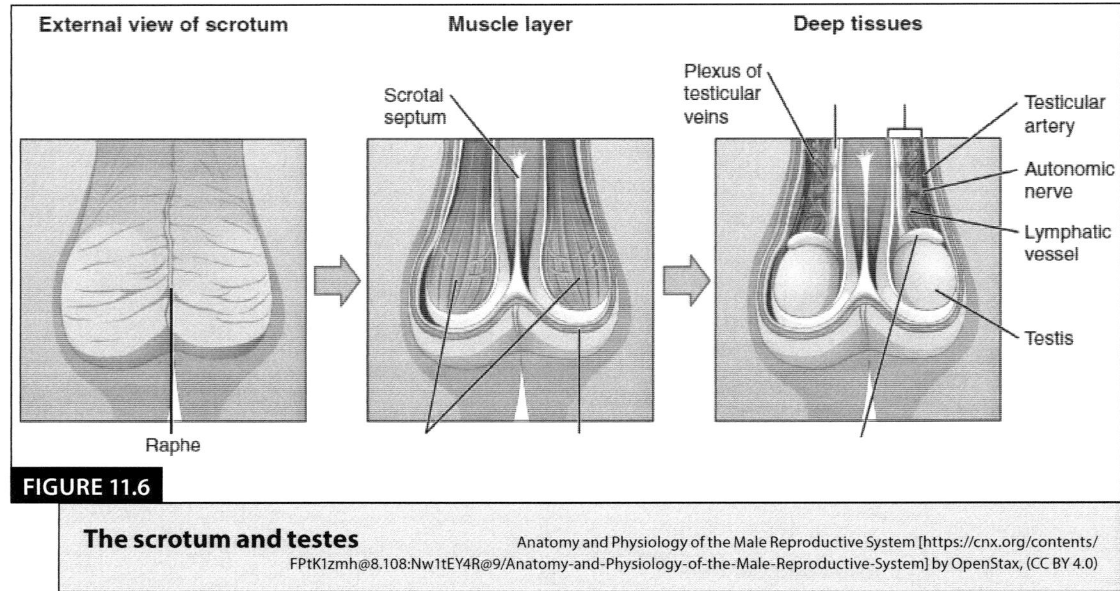


**FIGURE 11.6**

**The scrotum and testes**  Anatomy and Physiology of the Male Reproductive System [https://cnx.org/contents/ FPtK1zmh@8.108:Nw1tEY4R@9/Anatomy-and-Physiology-of-the-Male-Reproductive-System] by OpenStax, (CC BY 4.0)

2.  What is the **testicular temperature** for optimal sperm production? What structure(s) modify temperature? How do they modify it specifically?

3.  **Draw** in the space below a testis and include all of the following structures in your drawing: **cremaster muscle, dartos muscle, spermatic cord, scrotum, tunica vaginalis, tunica albuginea, seminiferous tubules, lobule, septum, straight tubules, rete testis, efferent ducts, ductus epididymis,** and **ductus deferens.**

    **Draw** here:

4.  Label the figure below once you have found these structures on pictures in your text or on the models in lab.

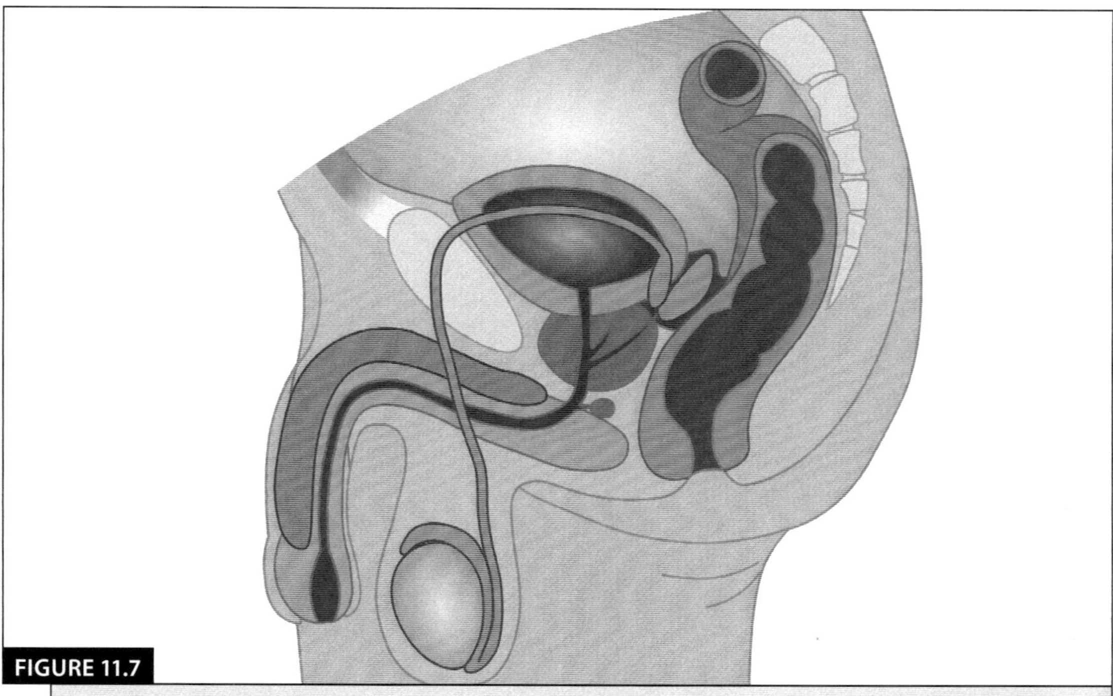

**FIGURE 11.7**

**Male reproductive system, midsagittal section.** Label glands and ducts of the male system, as well as the corpora cavernosa and spongiosum. This file is licensed under the Creative Commons Attribution-Share Alike 3.0. https://commons.wikimedia.org/wiki/File:Male_reproductive_system.png#globalusage

## ACTIVITY 2: EXAMINE THE PROCESSES OF SPERMATOGENESIS

*You will need a slide of sperm and a compound light microscope (if available) to complete this exercise.*

1. Construct your own flow chart detailing the process of spermatogenesis. Include in your chart the following terms: **spermatogonium, primary spermatocyte, secondary spermatocyte, spermatid, spermatozoa, mitosis, meiosis I, meiosis II,** and **spermiogenesis.**

2.  Briefly explain why mature gametes carry only one set of chromosomes.

_______________________________________________

_______________________________________________

_______________________________________________

_______________________________________________

_______________________________________________

3.  Sketch in the circle below a sperm cell when viewed under the compound light microscope. Label the following structures on your drawing below: **acrosome, nucleus, midpiece,** and **flagellum.**

4.  What special features are evident in sperm cells but not in somatic cells, and how do these specializations function?

_______________________________________________

_______________________________________________

_______________________________________________

_______________________________________________

_______________________________________________

5.  What do each of the three male accessory glands contribute to the semen?

## ACTIVITY 3: DISCUSS THE HORMONAL CONTROL OF THE MALE REPRODUCTIVE SYSTEM

*You will need a slide of sperm and a compound light microscope (if available) to complete this exercise.*

1.  Construct a flow chart detailing the control of testosterone. Include in your chart the following structures and hormones: **hypothalamus, adenohypophysis, testis, GnRH, FSH, LH, testosterone,** and **inhibin.**

2. While anabolic steroids (synthetic testosterone) bulk up muscles, they can also affect testosterone production in the testis. Using what you know about negative feedback, describe what would happen to testosterone production in the testis if a male takes large amounts of synthetic testosterone.

# 12

# FEMALE REPRODUCTIVE SYSTEM
## PRE-LAB

Name: _______________________    Section: __________    Date: _________

*Print this Lab handout to complete before attending your lab time, or fill in to return to your instructor or Lab TA prior to receiving the Lab handout.*

## LEARNING OBJECTIVES

1. Identify, locate, and describe the function of the structures in the female reproductive systems.

2. Examine the processes of oogenesis.

3. Discuss the hormonal control of the female reproductive system.

4. Describe the microscopic structure of the ovaries and uterus.

*Checklist to complete* **before entering** *the science skills lab (SSL):*

☐ Actively read this packet of information.

☐ Complete the charts, tables or labeling and answer questions using your own words.

☐ Complete the electronic digital pre-lab quiz on Bb by Sunday.

☐ Review the attached anatomy list and take it to lab with you. Jot down key descriptive identifying words that aid in your lab test preparations.

# ACTIVITY 1: IDENTIFY, LOCATE, AND DESCRIBE THE FUNCTION OF THE STRUCTURES IN THE FEMALE REPRODUCTIVE SYSTEMS

*Read the content in your text and the introductory paragraphs below. Then, complete Activity 1 in lab.*

The female reproductive system functions to produce gametes and reproductive hormones, just like the male reproductive system; however, it also has the additional task of supporting the developing fetus and delivering it to the outside world. Unlike its male counterpart, the female reproductive system is located primarily inside the pelvic cavity. Recall that the ovaries are the female gonads. The gamete they produce is called an **oocyte.** We will discuss the production of oocytes in detail shortly. First, let's look at some of the structures of the female reproductive system.

## VAGINA

The **vagina** is a muscular canal (approximately 10 cm long) that serves as the entrance to the reproductive tract. It also serves as the exit from the uterus during menses and childbirth. The outer walls of the anterior and posterior vagina are formed into longitudinal columns, or ridges, and the superior portion of the vagina—called the fornix—meets the protruding uterine cervix. The walls of the vagina are lined with an outer, fibrous adventitia, a middle layer of smooth muscle, and an inner mucous membrane with transverse folds called **rugae.** Rugae allows the expansion of the vagina to accommodate intercourse and childbirth.

The vagina is home to a normal population of microorganisms that help to protect against infection by pathogenic bacteria, yeast, or other organisms that can enter the vagina. In a healthy woman, the most predominant type of vaginal bacteria is from the genus *Lactobacillus.* This family of beneficial bacterial flora secretes lactic acid and thus, protects the vagina by maintaining an acidic pH (below 4.5). Potential pathogens are less likely to survive in these acidic conditions. Lactic acid, in combination with other vaginal secretions, makes the vagina a self-cleansing organ.

## OVARIES

The **ovaries** are the female gonads. Paired ovals, they are each about 2 to 3 cm in length, about the size of an almond. The ovaries are located within the pelvic cavity and are supported by the mesovarium, an extension of the peritoneum that connects the ovaries to the **broad ligament.** Extending from the mesovarium itself is the **suspensory ligament** that contains the ovarian blood and lymph vessels. Finally, the ovary itself is attached to the uterus via the **ovarian ligament.** The ovary is covered by dense connective tissue covering called the tunica albuginea. Beneath the tunica albuginea is the **cortex,** or outer portion, of the organ. Oocytes develop within the cortex, each surrounded by supporting cells. This grouping of an oocyte and its supporting cells is called a **follicle.** Beneath the cortex lies the inner ovarian medulla, the site of blood vessels, lymph vessels, and the nerves of the ovary.

**Label** the figures below.

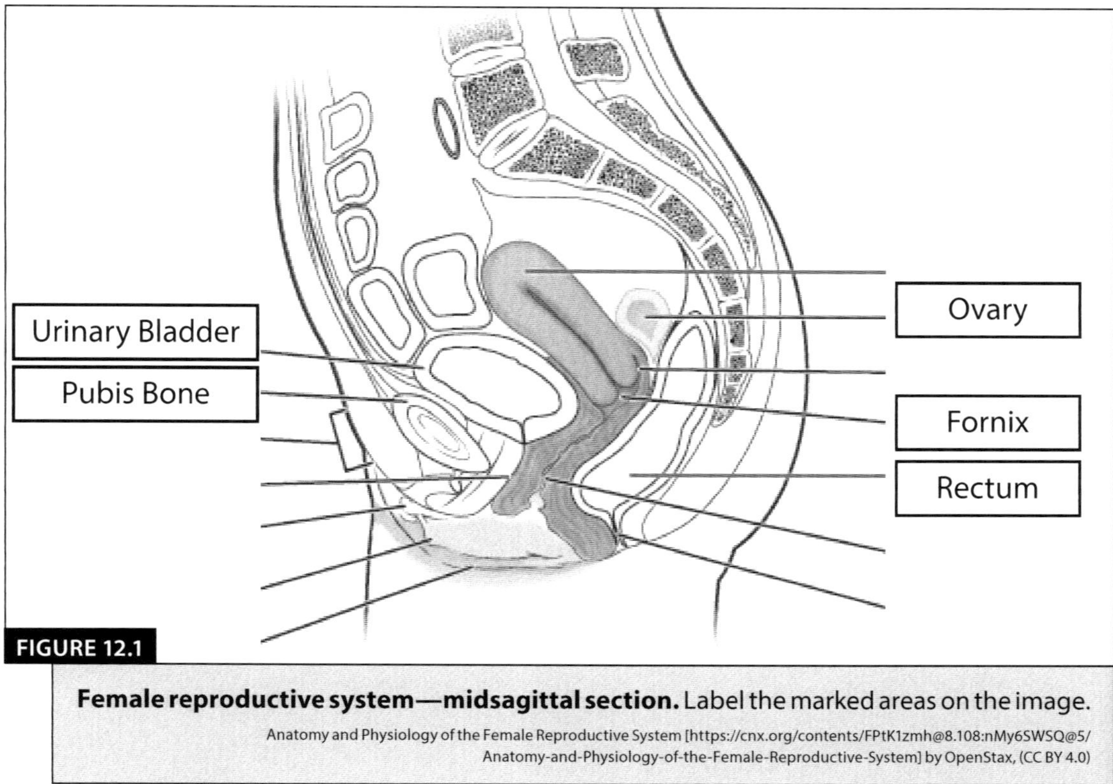


**FIGURE 12.1**

**Female reproductive system—midsagittal section.** Label the marked areas on the image.

Anatomy and Physiology of the Female Reproductive System [https://cnx.org/contents/FPtK1zmh@8.108:nMy6SWSQ@5/ Anatomy-and-Physiology-of-the-Female-Reproductive-System] by OpenStax, (CC BY 4.0)

## THE UTERINE TUBES

The **uterine tubes** (also called fallopian tubes or oviducts) serve as the conduit of the oocyte from the ovary to the uterus. Each of the two uterine tubes is close to, but not directly connected to, the ovary and divided into sections. The **isthmus** is the narrow medial end of each uterine tube that is connected to the uterus. The wide distal **infundibulum** flares out with slender, finger-like projections called **fimbriae.** The middle region of the tube, called the **ampulla,** is where fertilization often occurs. The uterine tubes also have three layers: an outer serosa, a middle smooth muscle layer, and an inner mucosal layer. In addition to its mucus-secreting cells, the inner mucosa contains ciliated cells that beat in the direction of the uterus, producing a current that will be critical to move the oocyte.

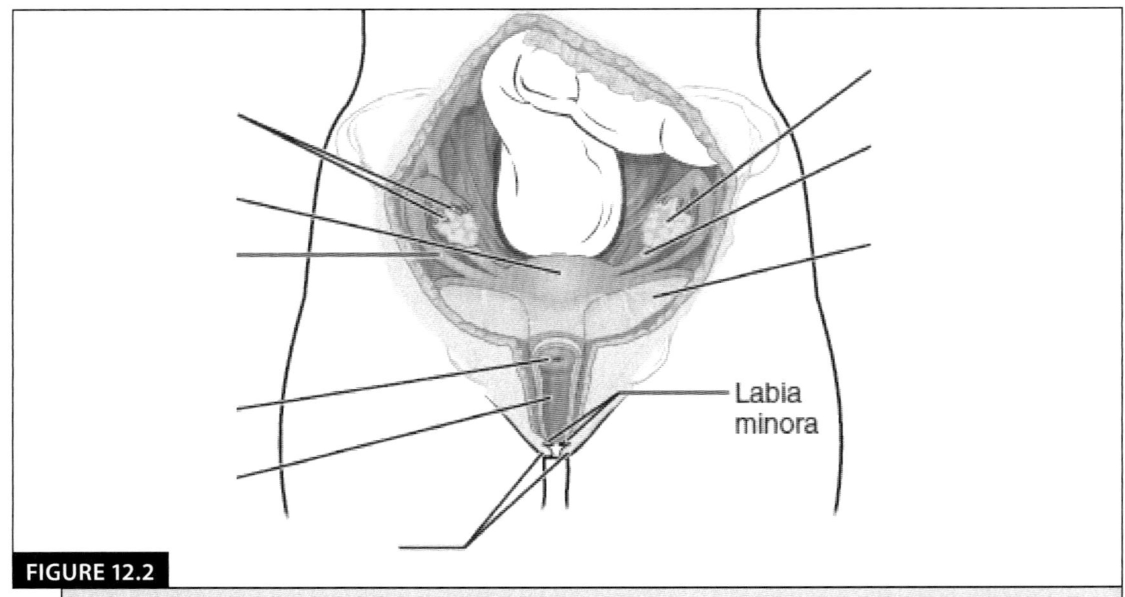


**FIGURE 12.2**

**Ovaries, uterine tubes, uterus and vagina–frontal or coronal section.** Label the structures indicated after reviewing the Pre-Lab, text, and lab models or diagrams.
Anatomy and Physiology of the Female Reproductive System [https://cnx.org/contents/FPtK1zmh@8.108:nMy6SWSQ@5/Anatomy-and-Physiology-of-the-Female-Reproductive-System] by OpenStax, (CC BY 4.0)

Following ovulation, the secondary oocyte surrounded by a few granulosa cells is released into the peritoneal cavity. The nearby uterine tube, either left or right, receives the oocyte. If the oocyte is successfully fertilized, the resulting zygote will begin to divide into two cells, then four, and so on, as it makes its way through the uterine tube and into the uterus. There, it will implant and continue to grow. If the egg is not fertilized, it will simply degrade—either in the uterine tube or in the uterus, where it may be shed with the next menstrual period.

## THE UTERUS AND CERVIX

The **uterus** is the muscular organ that nourishes and supports the growing embryo. Its average size is approximately 5 cm wide by 7 cm long (approximately 2 in by 3 in) when a female is not pregnant. It has three sections. The portion of the uterus superior to the opening of the uterine tubes is called the **fundus.** The middle section of the uterus is called the **body of uterus** (or corpus). The **cervix** is the narrow inferior portion of the uterus that projects into the vagina.

Several ligaments maintain the position of the uterus within the abdominopelvic cavity. The broad ligament is a fold of peritoneum that serves as a primary support for the uterus, extending laterally from both sides of the uterus and attaching it to the pelvic wall. The round ligament attaches to the uterus near the uterine tubes and extends to the labia majora. Finally, the uterosacral ligament stabilizes the uterus posteriorly by its connection from the cervix to the pelvic wall.

The wall of the uterus is made up of three layers. The most superficial layer is the serous membrane, or **perimetrium,** which consists of epithelial tissue that covers the exterior portion of the uterus. The middle layer, or **myometrium,** is a thick layer of smooth muscle responsible for uterine contractions. The innermost layer of the uterus that lines the lumen is called the **endometrium.** Structurally, the endometrium consists of two layers: the stratum basalis and the stratum functionalis. The stratum basalis layer is adjacent to the myometrium; this layer does not shed during menses. In contrast, the thicker stratum functionalis layer grows and thickens in response to increased levels of estrogen and progesterone. It is only the stratum functionalis layer of the endometrium that sheds during menstruation, or the **menses.** The first menses after puberty, called **menarche,** can occur either before or after the first ovulation.

**Label** the figure below.

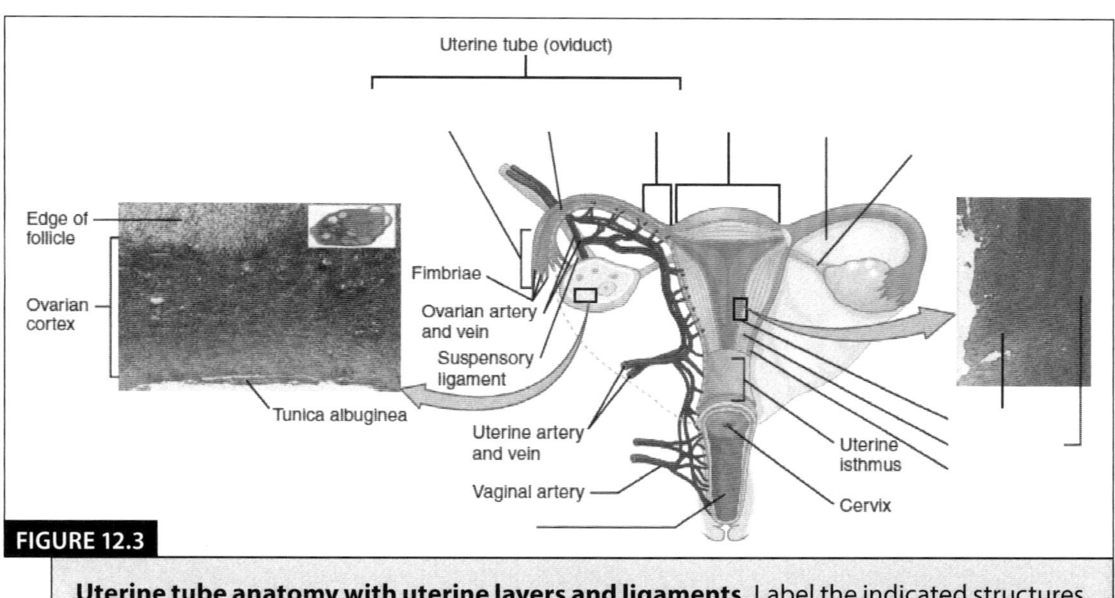


**FIGURE 12.3**

**Uterine tube anatomy with uterine layers and ligaments.** Label the indicated structures.

Anatomy and Physiology of the Female Reproductive System [https://cnx.org/contents/FPtK1zmh@8.108:nMy6SWSQ@5/ Anatomy-and-Physiology-of-the-Female-Reproductive-System] by OpenStax, (CC BY 4.0)

## EXTERNAL FEMALE GENITALS

The external female reproductive structures are referred to collectively as the **vulva.** The **mons pubis** is a pad of fat that is located at the anterior, over the pubic bone. After puberty, it becomes covered in pubic hair. The **labia majora** (labia = "lips;" majora = "larger") are folds of hair-covered skin that begin just posterior to the mons pubis. The thinner and more pigmented **labia minora** (labia = "lips;" minora = "smaller") extend medial to the labia majora. Although they naturally vary in shape and size from woman to woman, the labia minora serve to protect the female urethra and the entrance to the female reproductive tract.

The superior, anterior portions of the labia minora come together to encircle the **clitoris** (or glans clitoris), an organ that originates from the same cells as the glans penis and has abundant nerves that make it important in sexual sensation and orgasm. The **hymen** is a thin membrane that sometimes partially covers the entrance to the vagina. An intact hymen cannot be used as an indication of "virginity"; even at birth, this is only a partial membrane, as menstrual fluid and other secretions must be able to exit the body, regardless of penile–vaginal intercourse. The vaginal opening is located between the opening of the urethra and the anus. It is flanked by outlets to the **Bartholin's glands** (or greater vestibular glands). The Bartholin's glands and the lesser vestibular glands (located near the clitoris) secrete mucus, which keeps the vestibular area from drying out.

**Label** the figure below.

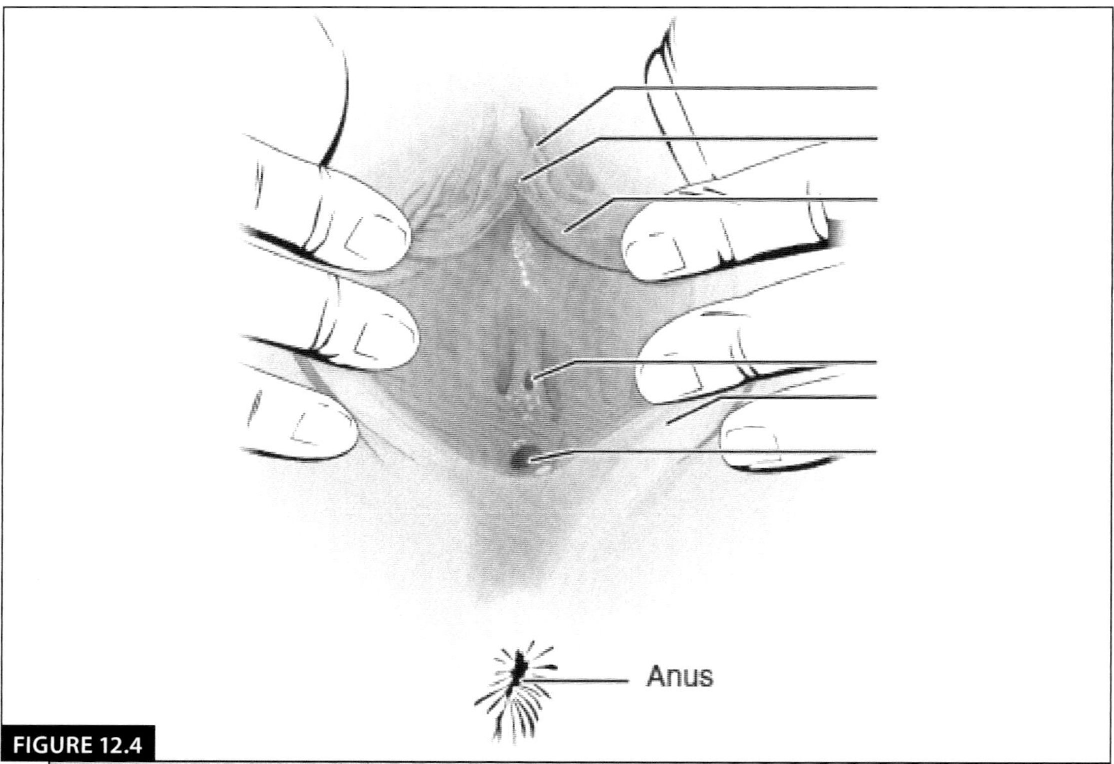


**FIGURE 12.4**

**The vulva–external anterior view.** Label the indicated structures.
Anatomy and Physiology of the Female Reproductive System [https://cnx.org/contents/ FPtK1zmh@8.108:nMy6SWSQ@5/Anatomy-and-Physiology-of-the-Female-Reproductive-System] by OpenStax, (CC BY 4.0)

## THE BREASTS

Even though the breasts are located far from the other female reproductive organs, they are considered accessory organs of the female reproductive system. The function of the breasts is to supply milk to an infant in a process called lactation. The external features of the breast include a nipple surrounded by a pigmented **areola.** The areola is typically circular and can vary in size from 25 to 100 mm in diameter.

Breast milk is produced by the **mammary glands,** which are modified sweat glands. The milk itself exits the breast through the nipple via 15 to 20 **lactiferous ducts** that open on the surface of the nipple. These lactiferous ducts each extend to a **lactiferous sinus** that connects to a glandular lobe within the breast itself that contains groups of milk-secreting cells in clusters called **alveoli.** The clusters can change in size depending on the amount of milk in the alveolar lumen. The lobes themselves are surrounded by fat tissue, which determines the size of the breast; breast size differs between individuals and does not affect the amount of milk produced. Supporting the breasts are multiple bands of connective tissue called **suspensory ligaments** that connect the breast tissue to the dermis of the overlying skin.

**Label** the figures below.

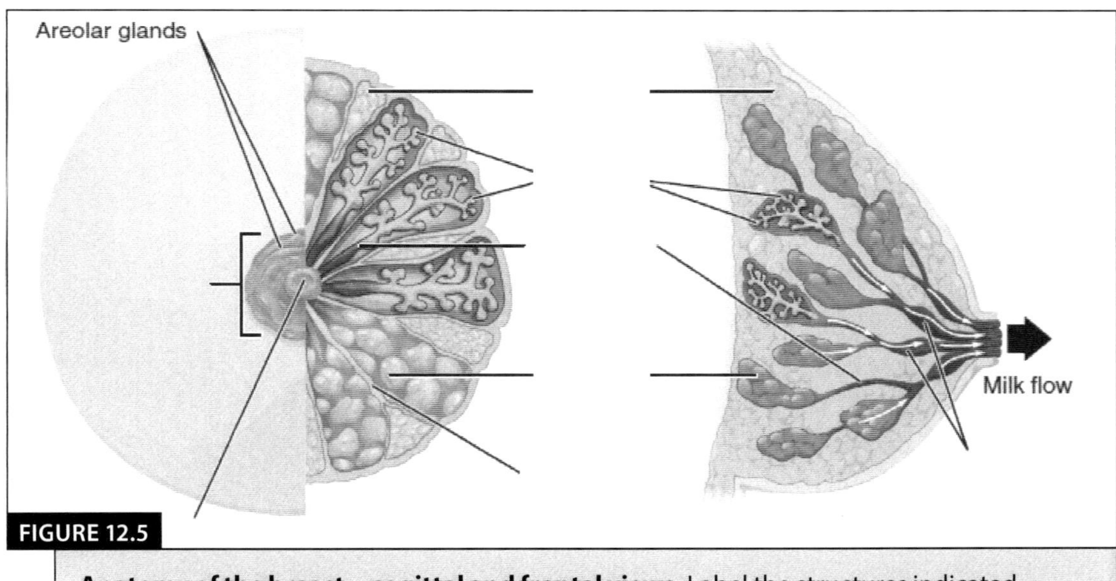


**FIGURE 12.5**

**Anatomy of the breast—sagittal and frontal views.** Label the structures indicated.
Anatomy and Physiology of the Female Reproductive System [https://cnx.org/contents/FPtK1zmh@8.108:nMy6SWSQ@5/Anatomy-and-Physiology-of-the-Female-Reproductive-System] by OpenStax, (CC BY 4.0)

## ACTIVITY 2: EXAMINE THE PROCESS OF OOGENESIS

*Read the content in your text and the introductory paragraphs below. Then, complete Activity 2 in lab.*

Gametogenesis in females is called **oogenesis.** The process begins with the ovarian stem cells, or **oogonia.** Oogonia are formed during fetal development, and divide via mitosis. However, oogonia form primary oocytes in the fetal ovary prior to birth. These primary oocytes are then arrested in this stage of meiosis I, only to resume it years later, beginning at puberty and continuing until the woman is near menopause (the cessation of a woman's reproductive functions).

The initiation of **ovulation**—the release of an oocyte from the ovary—marks the transition from puberty into reproductive maturity for women. From then on, throughout a woman's reproductive years, ovulation occurs approximately once every 28 days. Just prior to ovulation, a surge of luteinizing hormone triggers the resumption of meiosis in a primary oocyte. This initiates the transition from primary to secondary oocyte. However, this cell division does not result in two identical cells. Instead, the cytoplasm is divided unequally, and one daughter cell is much larger than the other. This larger cell, the secondary oocyte, eventually leaves the ovary during ovulation. The smaller cell, called the first **polar body,** may or may not complete meiosis and produce second polar bodies; in either case, it eventually disintegrates. Therefore, even though oogenesis produces up to four cells, only one survives.

How does the diploid secondary oocyte become an **ovum**—the haploid female gamete? Meiosis of a secondary oocyte is completed only if a sperm succeeds in penetrating its barriers. Meiosis II then resumes, producing one haploid ovum that, at the instant of fertilization by a (haploid) sperm, becomes the first diploid cell of the new offspring (a zygote). Thus, the ovum can be thought of as a brief, transitional, haploid stage between the diploid oocyte and diploid zygote. The larger amount of cytoplasm contained in the female gamete is used to supply the developing zygote with nutrients during the period between fertilization and implantation into the uterus.

Place the cells and processes correctly in the flow chart below.

1. Oogonium
2. Ovum
3. Primary ooocyte
4. Secondary oocyte
5. Mitosis
6. Meiosis I
7. Meiosis II
8. First polar body
9. Second polar body

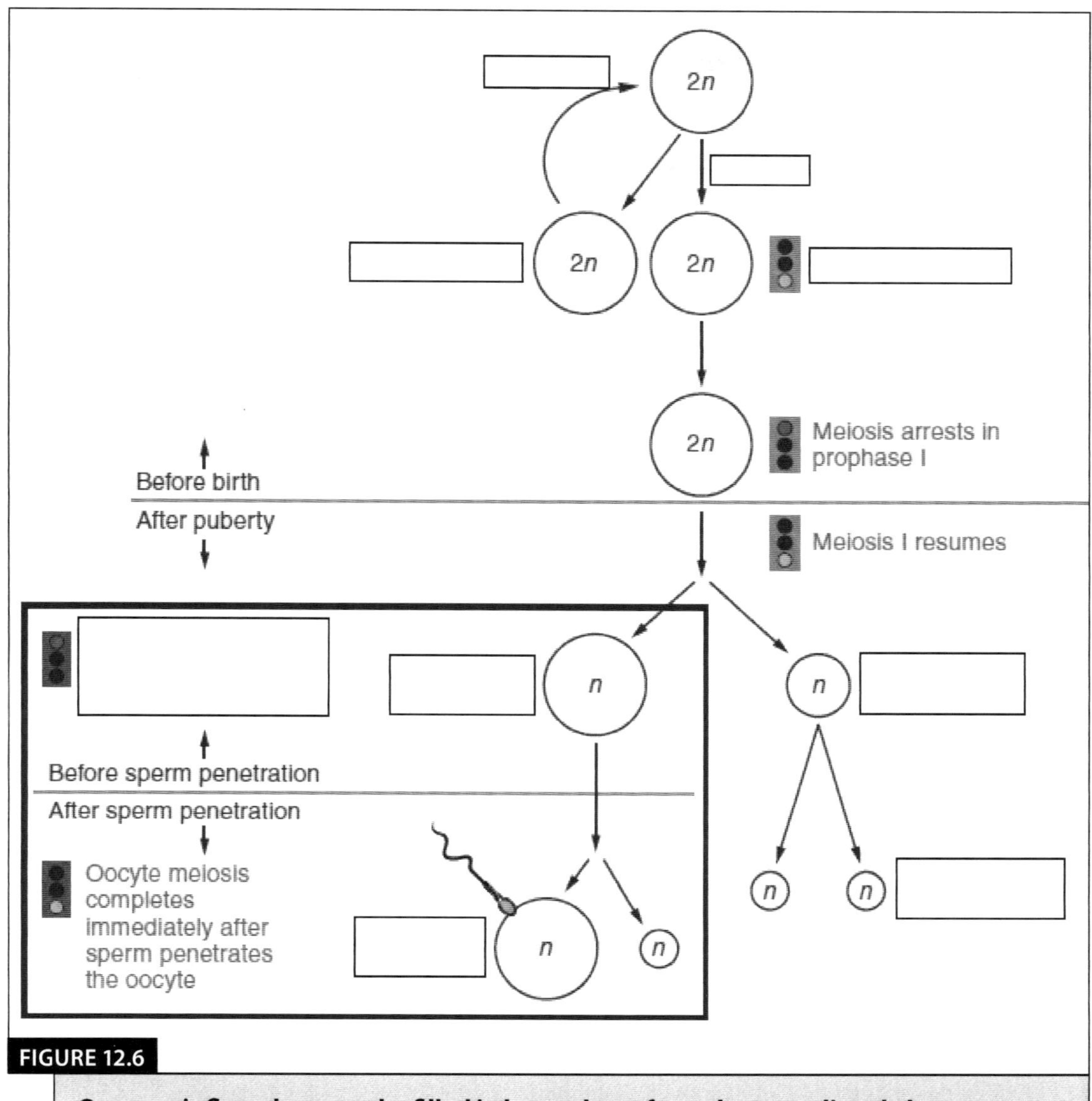


**FIGURE 12.6**

**Oogenesis flow chart—to be filled in by students from the terms listed above.** Anatomy and Physiology of the Female Reproductive System [https://cnx.org/contents/FPtK1zmh@8.108:nMy6SWSQ@5/Anatomy-and-Physiology-of-the-Female-Reproductive-System] by OpenStax, (CC BY 4.0)

# ACTIVITY 3: DISCUSS THE HORMONAL CONTROL OF THE FEMALE REPRODUCTIVE SYSTEM

*Read the content in your text and the introductory paragraphs below. Then, complete Activity 3 in lab.*

The process of oogenesis and folliculogenesis, from primordial follicle to early tertiary follicle, takes approximately two months in humans. The final stages of development end with ovulation of a secondary oocyte, which occurs over a course of approximately 28 days. These changes are regulated by many of the same hormones that regulate the male reproductive system, including GnRH, LH, and FSH.

**Read** your course text and complete the chart below.

**Hormones of oogenesis and folliculogenesis.** For each hormone listed, give its specific function in the process of oogenesis and/or folliculogenesis and / or the menstrual cycle.

**TABLE 12.1**

| HORMONE | FUNCTION |
|---|---|
| GnRH | |
| LH | |
| FSH | |
| Estrogen/ Estradiol | |
| Progesterone | |
| Inhibin | |
| Relaxin | |

# ACTIVITY 4: DESCRIBE THE MICROSCOPIC STRUCTURE OF THE OVARIES AND UTERUS

*Read the content in your text and the introductory paragraphs below. Then, complete Activity 4 in lab.*

Again, ovarian follicles are oocytes and their supporting cells. They grow and develop in a process called **folliculogenesis,** which typically leads to ovulation of one follicle approximately every 28 days, along with death to multiple other follicles. Recall that a female infant at birth will have one to two million oocytes within her ovarian follicles, and that this number declines throughout life until menopause, when no follicles remain.

Folliculogenesis begins with follicles in a resting state. These small **primordial follicles** are present in newborn females and are the prevailing follicle type in the adult ovary. Primordial follicles have only a single flat layer of support cells, called **granulosa cells,** that surround the oocyte. After puberty, a few primordial follicles will respond to a recruitment signal each day and will join a pool of immature growing follicles called **primary follicles.** Primary follicles start with a single layer of granulosa cells, but the granulosa cells then become active and transition from squamous shape to a cuboidal shape as they increase in size and proliferate. As the granulosa cells divide, the follicles—now called **secondary follicles**—increase in diameter, adding a new outer layer of connective tissue, blood vessels, and **theca cells**—cells that work with the granulosa cells to produce estrogens.

Within the growing secondary follicle, the primary oocyte now secretes a thin acellular membrane called the zona pellucida that will play a critical role in fertilization. A thick fluid, called follicular fluid, that has formed between the granulosa cells also begins to collect into one large pool, or **antrum.** Follicles in which the antrum has become large and fully formed are considered **tertiary follicles** (or antral follicles). Several follicles reach the tertiary stage at the same time, and most of these will undergo atresia. The one that does not die will continue to grow and develop until ovulation, when it will expel its secondary oocyte surrounded by several layers of granulosa cells from the ovary.

Label the primary structures of each stage of the figure below.

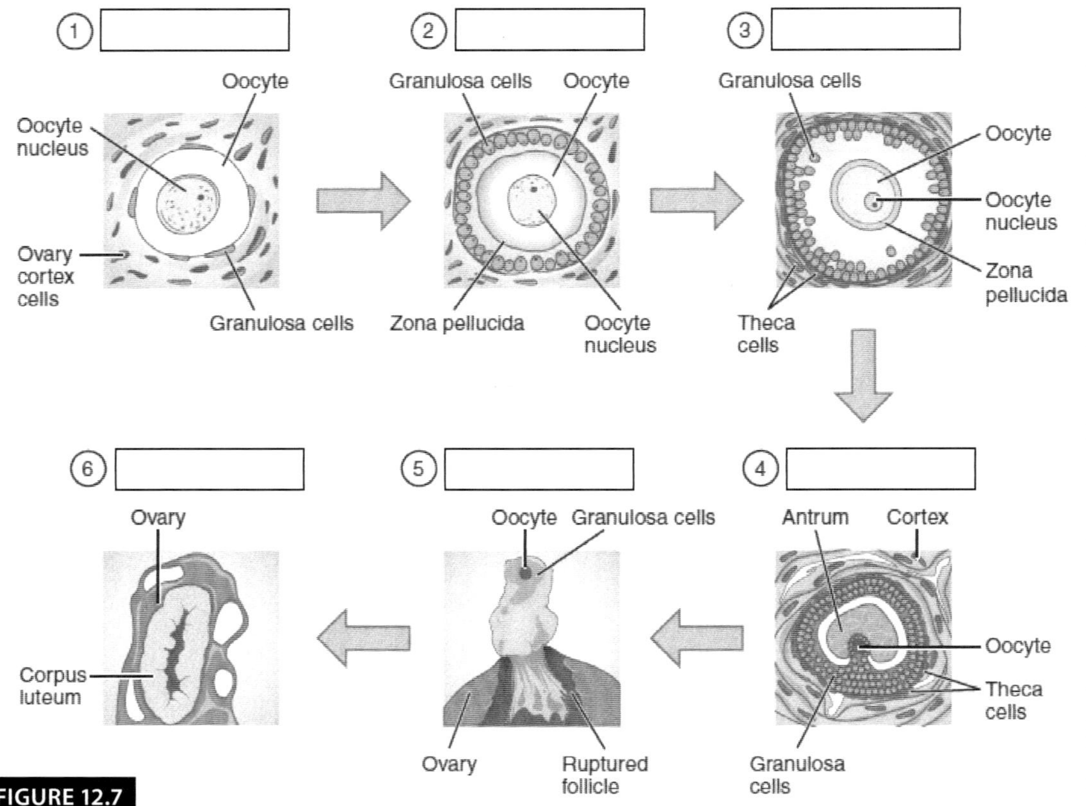

**FIGURE 12.7**

**Folliculogenesis**    Anatomy and Physiology of the Female Reproductive System [https://cnx.org/contents/FPtK1zmh@8.108:nMy6SWSQ@5/Anatomy-and-Physiology-of-the-Female-Reproductive-System] by OpenStax, (CC BY 4.0)

# THE MENSTRUAL CYCLE

The series of changes in which the uterine lining is shed, rebuilds, and prepares for implantation in known as the menstrual cycle. The timing of the menstrual cycle starts with the first day of menses, referred to as day one of a woman's period. The average length of a woman's menstrual cycle is 28 days; this is the time period used to identify the timing of events in the cycle. However, the length of the menstrual cycle varies among women, and even in the same woman from one cycle to the next, typically from 21 to 32 days. Just as the hormones produced by the granulosa and theca cells of the ovary "drive" the follicular and luteal phases of the ovarian cycle, they also control the three distinct phases of the menstrual cycle. These are the menses phase, the proliferative phase, and the secretory phase.

## MENSES PHASE

The **menses phase** of the menstrual cycle is the phase during which the lining is shed; that is, the days that the woman menstruates. Although it averages approximately five days, the menses phase can last from 2 to 7 days, or longer. The menses phase occurs during the early days of the follicular phase of the ovarian cycle, when progesterone, FSH, and LH levels are low. Recall that progesterone concentrations decline as a result of the degradation of the corpus luteum, marking the end of the luteal phase. This decline in progesterone triggers the shedding of the stratum functionalis of the endometrium.

## PROLIFERATIVE PHASE

Once menstrual flow ceases, the endometrium begins to proliferate again, marking the beginning of the **proliferative phase** of the menstrual cycle. It occurs when the granulosa and theca cells of the tertiary follicles begin to produce increased amounts of estrogen. These rising estrogen concentrations stimulate the endometrial lining to rebuild. In a typical 28-day menstrual cycle, ovulation occurs on day 14. Ovulation marks the end of the proliferative phase as well as the end of the follicular phase.

## SECRETORY PHASE

In the ovary, the luteinization of the granulosa cells of the collapsed follicle forms the progesterone producing corpus luteum, marking the beginning of the luteal phase of the ovarian cycle. In the uterus, progesterone from the corpus luteum begins the **secretory phase** of the menstrual cycle, in which the endometrial lining prepares for implantation. Over the next 10 to 12 days, the endometrial glands secrete a fluid rich in glycogen. If fertilization has occurred, this fluid will nourish the ball of cells now developing from the zygote. At the same time, the spiral arteries develop to provide blood to the thickened stratum functionalis.

If no pregnancy occurs within approximately 10 to 12 days, the corpus luteum will degrade into the corpus albicans. Levels of both estrogen and progesterone will fall, and the endometrium will grow thinner. Prostaglandins will be secreted that cause constriction of the spiral arteries, reducing oxygen supply. The endometrial tissue will die, resulting in menses—or the first day of the next cycle.

*Checklist to complete* **before entering** *the science skills lab (SSL):*

- ☐ Actively read this packet of information.
- ☐ Complete the charts, tables or labeling and answer questions using your own words.
- ☐ Complete the electronic digital pre-lab quiz on Bb by Sunday.
- ☐ Review the attached anatomy list and take it to lab with you. Jot down key descriptive identifying words that aid in your lab test preparations.

## ANATOMY LIST

1. Identify these structures of the **female reproductive system** on the appropriate models or diagrams in lab:

   **a.** ovary

   **b.** fimbriae

   **c.** uterine tube (fallopian tube) (add three parts–optional)

   **d.** ovarian ligament

   **e.** uterus

      **i.** fundus

      **ii.** body

      **iii.** cervix

         1. cervical os (external and internal)

      **iv.** myometrium

      **v.** endometrium

   **f.** broad ligament

   **g.** round ligament

   **h.** vagina

      **i.** fornix

      **ii.** vaginal canal

   **i.** external genitalia (vulva)

      **i.** mons pubis

      **ii.** labia majora

      **iii.** labia minora

      **iv.** vestibule

         1. Vestibular glands (Bartholin's glands)

      **v.** hymen

      **vi.** glans of the clitoris

      **vii.** perineum

         1. anterior triangle (urogenital triangle)

         2. posterior triangle (anal triangle)

         3. obstetric perineum (between anus and birth canal/vagina)

    **j.**  mammary glands

        **i.**  mammary glands

        **ii.**  nipple

        **iii.**  areola

**2.**  Identify the following microscopic structures on a **slide** of the **ovary:**

    **a.**  primordial follicles (containing primary oocytes)

    **b.**  Graafian follicle

        **i.**  follicular cells (granulosa cells)

        **ii.**  thecal cells (excessive proliferation = infertility)

        **iii.**  oocyte

        **iv.**  zona pellucida

        **v.**  corona radiata

    **c.**  corpus luteum

*Bring this lab list to the lab with you this week. Remember to show this pre-lab to the SSL instructor **before entering** the lab. It will be marked off in the SSL book for grade credit towards your lab test.*

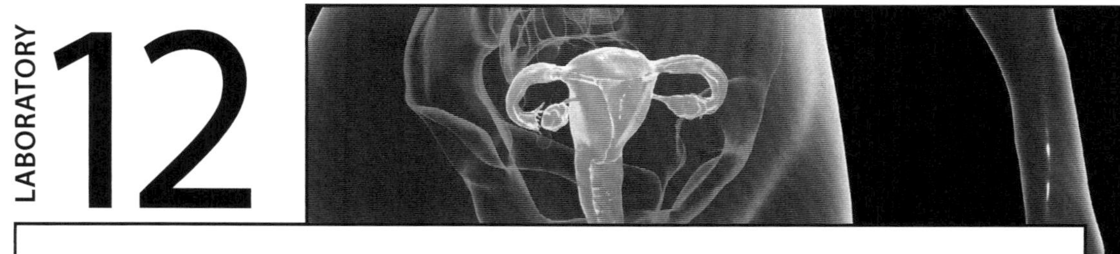

# FEMALE REPRODUCTIVE SYSTEM
## IN-LAB ACTIVITIES

Name: _______________________    Section: ___________    Date: _________

## ACTIVITY 1: IDENTIFY, LOCATE, AND DESCRIBE THE FUNCTION OF THE STRUCTURES IN THE FEMALE REPRODUCTIVE SYSTEMS

*You will need female reproductive models (if available) to complete this exercise.*

1. Do your best to draw the following below from the models in lab (in the anterior view): **ampulla, fimbrae, infundibulum, ovary,** and **uterine tube.**

2. Label the figure below.

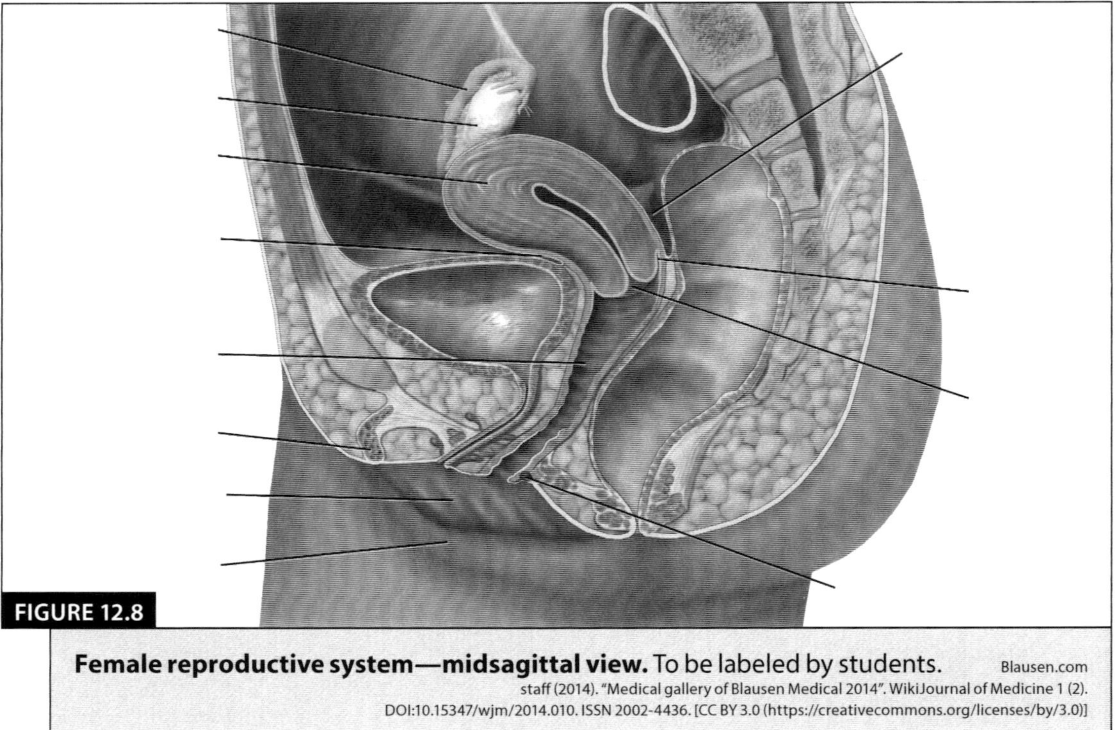

**FIGURE 12.8**

**Female reproductive system—midsagittal view.** To be labeled by students. Blausen.com staff (2014). "Medical gallery of Blausen Medical 2014". WikiJournal of Medicine 1 (2). DOI:10.15347/wjm/2014.010. ISSN 2002-4436. [CC BY 3.0 (https://creativecommons.org/licenses/by/3.0)]

3. On the figure above, add in new arrows to identify the five additional structures that are found on your anatomy list. (**fimbrae, suspensory ligament, ovarian ligament, round ligament,** and **mons pubis**)

4. Do your best to draw the following below: **anus, clitoris, hymen, labia majora, mons pubis,** and **prepuce.** *Use the models in lab to help you.*

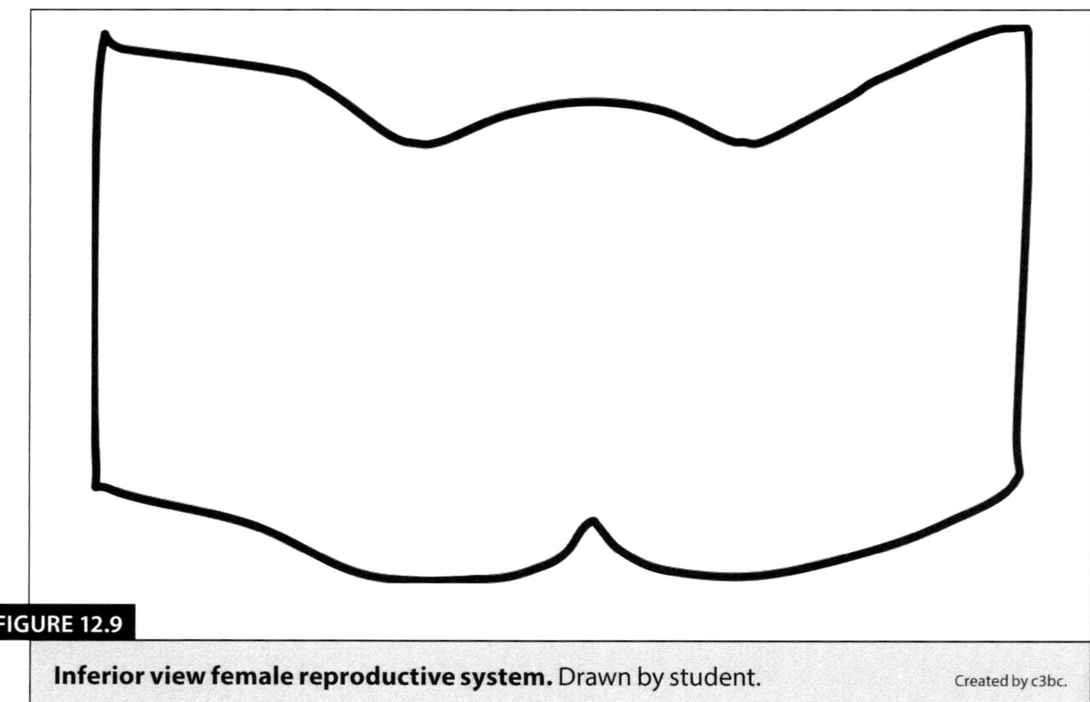

**FIGURE 12.9**

**Inferior view female reproductive system.** Drawn by student. Created by c3bc.

**5.** Do your best to draw the following below: **adipose tissue, areola, lactiferous duct, mammary duct,** and **nipple.**

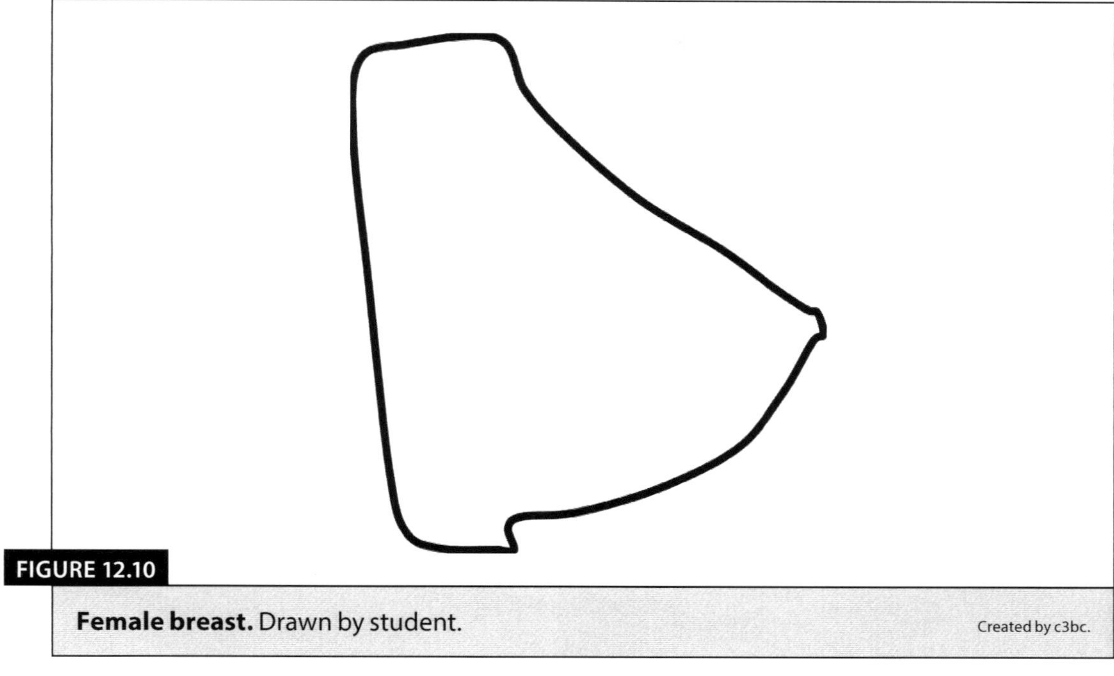

**FIGURE 12.10**

**Female breast.** Drawn by student.

Created by c3bc.

**6.** What makes the vagina a self-cleaning organ?

**7.** Follow the path of ejaculated sperm from the vagina to the oocyte. Include all structures of the female reproductive tract that the sperm must swim through to reach the egg.

8. Endometriosis is a disorder in which endometrial cells implant and proliferate outside of the uterus—in the uterine tubes, on the ovaries, or even in the pelvic cavity. Offer a theory as to why endometriosis increases a woman's risk of infertility.

## ACTIVITY 2: EXAMINE THE PROCESSES OF OOGENESIS

1. Construct your own flow chart detailing the process of oogenesis. Include in your chart the following terms: oogonium, primary oocyte, primary polar body, secondary oocyte, secondary polar bodies, ovum, ovulation, fertilization, mitosis, meiosis I, and meiosis II. Denote in your drawing which events happen before birth, from birth to puberty, and from puberty to fertilization.

2. Briefly explain polar bodies. Include in your answer their contents and significance.

3. Identify some differences between meiosis in men and women.

## ACTIVITY 3: DISCUSS THE HORMONAL CONTROL OF THE FEMALE REPRODUCTIVE SYSTEM

1. Construct a diagram that shows how the following hormones interact to facilitate oogenesis, folliculogenesis, and the menstrual cycle. You must include: **GnRH, LH, FSH, estrogen, progesterone,** and **inhibin** in your diagram.

2. Discuss the corpus luteum. Include in your answer: what it is, how it is made, what it does, and when it degrades.

3. Discuss the corpus albicans. Include in your answer: what it is, how it is made, what it does, and when it degrades.

4. How do hormonal birth control pills work?

## ACTIVITY 4: DESCRIBE THE MICROSCOPIC STRUCTURE OF THE OVARIES AND UTERUS

*You will need female reproductive models, ovary histology slide, and a compound light microscope (if available) to complete this exercise.*

1. Label the model below including all structures on your anatomy list that are visible.

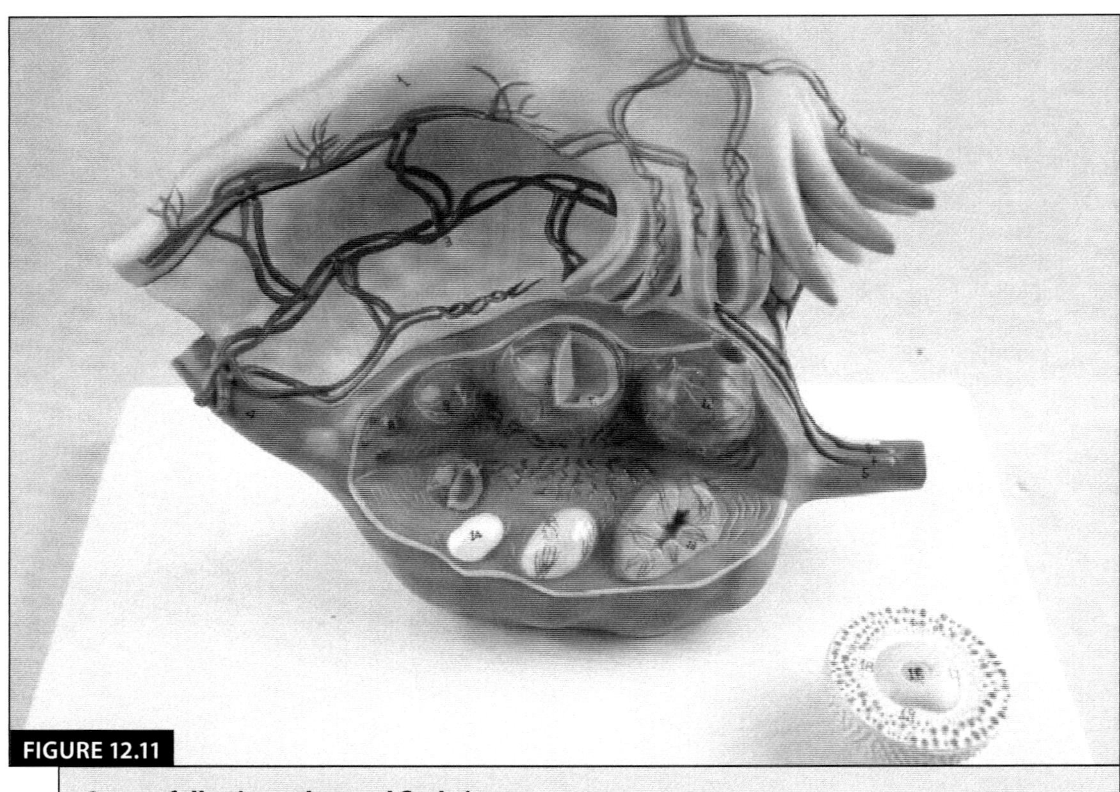

**FIGURE 12.11**

**Ovary, fallopian tube, and fimbriae**

2. Using the ovary histology slide and the compound light microscope, sketch in the circles below a **primordial follicle** (first circle); a **Graafian follicle**—identifying follicular cells (granulosa cells), thecal cells, oocyte, zona pellucida, and the corona radiate (all in the second circle); and a **corpus luteum** (third circle).

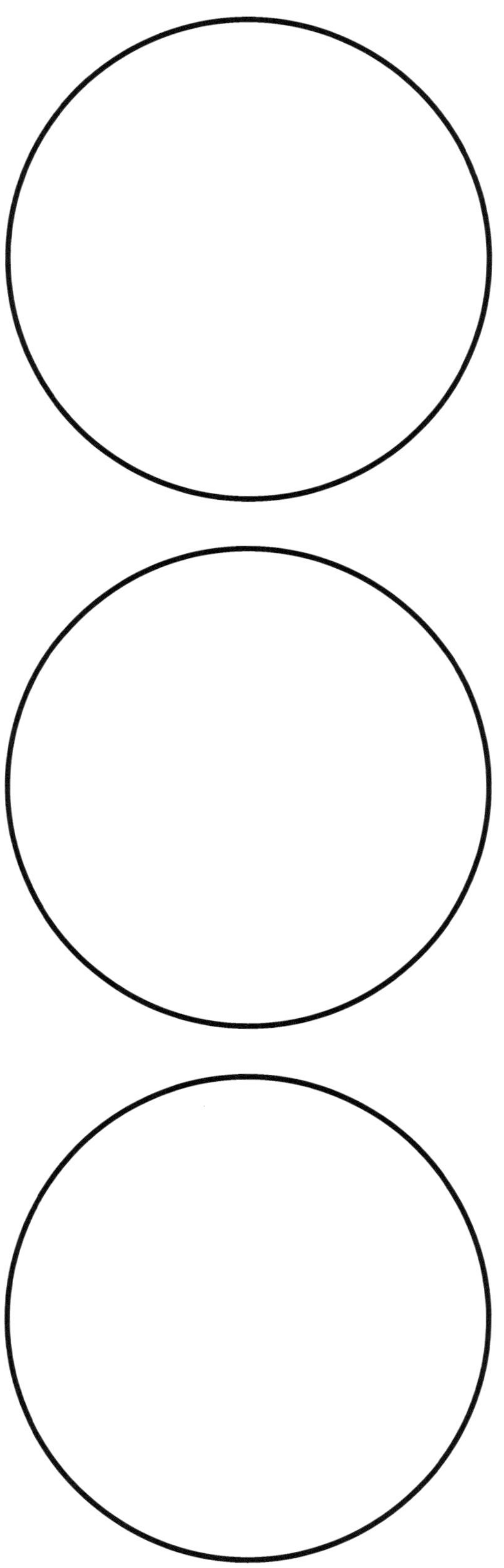

# 13

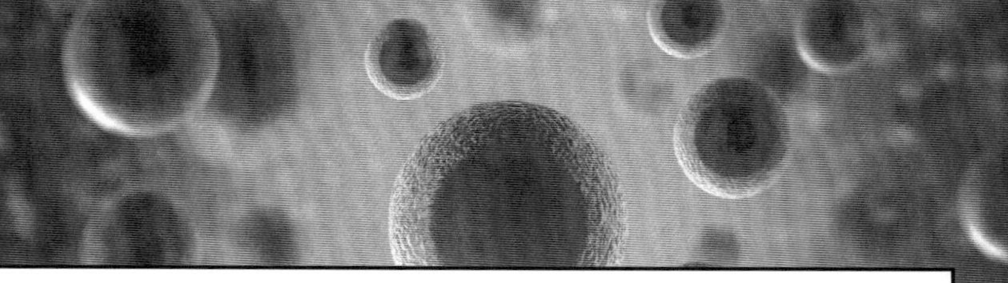

# DEVELOPMENT
## PRE-LAB

Name: _________________________ Section: __________ Date: _________

## LEARNING OBJECTIVES

1. Describe the three major pre-implantation forms of the human embryo: zygote, morula, and blastula (inner cell mass, blastocyst cavity, and trophoblast), including the post-fertilization time periods during which they exist.

2. Describe the ectoderm, mesoderm, and endoderm primary germ layers, including their origin and contributions to the amnion, chorion, and yolk sac. Describe at least two adult tissues that form from the differentiation of each germ layer.

3. Describe the contributions of the embryo and the mother to formation of the placenta. Describe the structure and processes by which the maternal and fetal circulation exchange components such as oxygen, carbon dioxide, nutrients, and wastes. What components are normally not exchanged?

4. Define the embryonic period in terms of post-fertilization time period, structures, and functions observed at the end of this period.

5. Describe the changes during fetal development, including the increased demands placed on maternal systems. In the lab activities, you will observe various stages of fetal development, as well as the differences in fraternal and identical twins.

6. Describe labor and parturition in terms of onset, phases, hormones, and physiological changes of the newborn.

*Checklist to complete* **before entering** *the science skills lab (SSL):*

☐ Actively read this packet of information.

☐ Complete the charts, tables or labeling and answer questions using your own words.

☐ Complete the electronic digital pre-lab quiz on Bb by Sunday.

☐ Review the attached anatomy list and take it to lab with you. Jot down key descriptive identifying words that aid in your lab test preparations.

# ACTIVITY 1:

*Read the content in your text and the introductory paragraphs below. Then, complete Activity 1 in lab.*

1.  Fertilization marks the onset of embryonic period that continues for another eight weeks. The **zygote** is a single cell that contains the combined genomic DNA of a sperm and ovum cell (gametes). The zygote forms after a sperm releases digestive enzymes from its acrosome, moves past cells bound to the outside of the secondary oocyte (corona radiata), past the glycoprotein layer surrounding the oocyte called the zona pellucida, and fuses its plasma membrane with that of the oocyte. The fusion of additional sperm with the oocyte are prevented by changes in membrane voltage and chemical changes in the plasma membrane of the oocyte and in the zona pellucida. These are called the fast and slow blocks to polyspermy, respectively. Meiosis is completed immediately after fertilization, making the secondary oocyte an ovum, briefly, until it becomes the zygote when its nucleus becomes one with the nucleus of the sperm. The ovum's contribution to the zygote includes maternal genomic DNA, most of the cytoplasm (including mitochondria), and most of the plasma membrane. The primary contribution of the sperm is paternal genomic DNA. The zygote normally has 46 chromosomes, whereas the sperm and ovum normally have 23 chromosomes each (i.e., Gametes are haploid cells having half the number of chromosomes of other cells.)  A 23-chromosome set from each of the parent's 46 chromosomes combine during fertilitation. This process is one that results in offspring with unique combinations of genetic information.

The secondary oocyte lives for approximately one day unless fertilization takes place. In other words, fertilization normally must take place within one day after ovulation. However, intercourse 3–5 days before ovulation normally can result in fertilization because sperm can live approximately 3-5 days in the female reproductive tract. Because the secondary oocyte is transported along the uterine tube/oviduct slowly by cilia, fertilization must usually take place in the first portion of the uterine tube/oviduct; the secondary oocyte does not live long enough to be transported further.

*Complete the summary table on fertilization below.*

| CELL TYPE | GENOMIC MAKE-UP (NUMBER AND ORIGIN) | LIFE SPAN IN FEMALE REPRODUCTIVE TRACT | ACTIVITIES | CONTRIBUTION TO ZYGOTE |
|---|---|---|---|---|
| Secondary Oocyte | 23 duplicated chromosomes of mother | | | Cytoplasm including mitochondria, maternal DNA |
| Sperm | | | Motility, acrosome reaction, binds and fuses with oocyte plasma membrane | |
| Ovum | 23 single chromosomes of mother | Short, only occurs if oocyte is fertilized | | |
| Zygote | | ~24 hr | | |

# ACTIVITY 2: CHARACTERIZE THE PRE-IMPLANTATION FORMS OF THE HUMAN EMBRYO

*Read the content in your text and the introductory paragraphs below. Then, complete Activity 2 in lab.*

1. The zygote begins to divide after 24 hours at an increasingly fast rate with little increase in the synthesis of new cellular materials. The new cells get smaller with each division leading to these cell divisions being called reduction divisions. A ball of cells, called a **morula,** is formed within the confines of the zona pellucida and is found at day 5 in the distal oviduct. This morula develops a fluid-filled cavity to become a **blastula,** and the zona pellucida is lost in preparation for implanting in the uterine wall, found at day 6 to 7 in the uterus. A cluster of cells forms at a single location on the inner surface of the blastula wall; it is called the embryoblast, and it will become the embryo and contribute to formation of embryonic membranes. The blastula wall cells are the trophoblast, and they contribute to formation of the chorionic membrane and placenta.

*Complete the summary table on pre-implantation embryonic forms below.*

| EMBRYONIC STAGE | POST-FERTILIZATION DATE OBSERVED | LOCATION FOUND | CELLULAR ORGANIZATION |
|---|---|---|---|
| Zygote | Day 1 | Proximal oviduct | Single large cell |
| Morula | | | Ball of cells, each much smaller than zygote |
| Blastula | | | |

2. Embryonic period. The embryonic period continues through the $8^{th}$ week post-fertilization. A major early event is formation of the three primary germ layers. The embryo at this stage is a relatively flat oval plate with three distinct shallow layers. The superficial layers are the **endoderm** and the **ectoderm;** developing between them is the **mesoderm** layer. The ectoderm develops into external surface structures including the epidermis and into the nervous system. The endoderm forms internal surface structures including the respiratory epithelium and the gastrointestinal tract. The mesodermal derivatives include muscle and bone. By the end of the embryonic period, the rudiments of the body systems are in place although they may not be functional.

3. Embryonic and fetal membranes. Four membranes are formed during embryonic development: chorion, amnion, yolk sac, and allantois. The chorion is derived from the trophoblast and interacts with the endometrium of the uterus to form the placenta and surround the embryo and all other membranes. The chorion is most developed near the embryo with the trophoblast cells invading the adjacent endometrium to form projections called chorionic villi that are highly vascularized by the embryo. These villi project into blood sinuses containing maternal blood and allow for exchange between the fetal and maternal circulatory systems. CO2 and O2 exchange by diffusion as do many small molecules. Most larger molecules are not exchanged, including antibodies, except for antibodies of the IgG class (which can cross the placenta). Importantly, rejection of the fetus by the maternal cellular immune system does not usually occur because cells are not normally exchanged across the placenta. Antibodies to the Rh antigen found on red blood cells are often IgG class, making rejection of a Rh antigen positive fetus by a Rh-negative mother a possibility. Many potentially harmful substances such as cocaine, alcohol, excess vitamins, and many prescription drugs can exchange with the fetal circulation. The **amnion** arises from the margins of the ectoderm to line the inside of the **chorion,** eventually surrounding the fetus save for the umbilical cord with the amniotic fluid that it contains. This allows the fetus to develop in a practically weightless environment preventing mechanical deformation and damage to the fetus. The **yolk sac** forms in a similar manner on the endodermal side of the embryo. In some animals, it is a significant source of energy stores but not so in humans. It does not grow with the embryo, so it quickly becomes small in comparison to the fetus. The **allantois** is an outgrowth of the developing gut that contributes to formation of the umbilical cord.

4. Fetal Development. This lasts from the $9^{th}$ through the $37^{th}$ week of development. During this period, the fetus grows dramatically in size and weight. The rudimentary systems developed during embryogenesis continue to grow and differentiate towards full functionality. The mother must supply all the energy, plus the materials that can't be synthesized by the fetus. This puts significantly higher demands on the mother's digestive, urinary, circulatory, respiratory, and skeletal/muscular systems. A rough estimate of the increased demands on these systems is 30% to 50%.

# ACTIVITY 3: CHARACTERIZE THE THREE STAGES OF LABOR  −

*Read the content in your text and the introductory paragraphs below. Then, complete Activity 3 in lab.*

1. Labor. The onset of labor can't be predicted with precision. **Progesterone** suppresses uterine contractions. It is thought that decreasing levels of progesterone while estrogen levels continue to rise near the end of the third trimester, leads to increasing uterine contractions. At the same time, oxytocin levels are rising due to secretion from the maternal and fetal pituitary glands. **Oxytocin** is a potent stimulator of smooth muscle contraction and prostaglandin synthesis, which also stimulates smooth muscle contraction. Increased contractions of the uterus begin to expel the fetus causing the cervix to stretch. This initiates a positive feedback mechanism involving stretch receptors in the cervix stimulating oxytocin release, further stimulating uterine contraction and stretching of the cervix. The cycle does not end until the fetus passes through the cervix and it ceases to stretch. The three phases of labor are **dilation, expulsion,** and **placental** ejection. The dilation phase refers to dilation of the cervix to a diameter greater than 10 cm. It usually lasts 6-12 hours but can vary dramatically from minutes to days. The expulsion phase refers to expelling the fetus though the cervical canal to the outside world, which takes less than two hours but can vary significantly. The normal position of the fetus during delivery is the vertex position (head first). Having the fetus in a breech position (legs first) can greatly complicate the delivery to the point of opting for surgery (C-section). The mother expels the placenta (afterbirth) in the final stage of labor as contractions of the uterus continue and weaken its attachment to the uterus. This final stage should occur within 30 minutes, or the placenta may need to be removed by mechanical force or surgery.

2. Complete the summary table on labor below.

| LABOR STAGE | DESCRIPTION | DURATION |
|---|---|---|
| Dilation | Onset of labor marked by uterine contractions and dilation of cervix to >10 cm. | 6–12 hr average but highly variable (minutes to days) |
| Expulsion |  |  |
| Placental |  |  |

*Checklist to complete* **before entering** *the science skills lab (SSL):*

☐ Actively read this packet of information.

☐ Complete the charts, tables or labeling and answer questions using your own words.

☐ Complete the electronic digital pre-lab quiz on Bb by Sunday.

☐ Review the attached anatomy list and take it to lab with you. Jot down key descriptive identifying words that aid in your lab test preparations.

## ANATOMY LIST—REVIEW PRIOR TO LAB

Describe and be able to identify the following cells and embryonic or fetal structures on **models** in the lab. "Fxn" indicates to include function with the description.

1. Secondary oocyte

2. Ovum

3. Sperm (fxn)

4. Corona radiata

5. Zona pellucida

6. Zygote

7. Morula

8. Blastocyst (fxn)

9. Inner cell mass

10. Trophoblast (fxn)

11. Amnion (fxn)

12. Chorion (fxn)

13. Yolk sac

14. Placenta (fxn)

15. Umbilical cord, including arteries and veins (fxn)

16. Chorionic villi (fxn)

*Remember to show this pre-lab to the SSL instructor* **before entering** *the lab. It will be marked off in the SSL book for grade credit towards your lab test.*

# 13

# DEVELOPMENT
## IN-LAB ACTIVITIES

Name: _________________________     Section: __________     Date: _________

## LEARNING OBJECTIVES

1. Describe the three major pre-implantation forms of the human embryo: zygote, morula, and blastula (inner cell mass, blastocyst cavity, and trophoblast), including the post-fertilization time periods during which they exist.

2. Examine the starfish development and compare it to mammalian development.

3. Using the chick embryo slides available, observe the amnion, chorion, and yolk sac. Locate and label five developmental parts of the chick embryo.

4. Define the embryonic period in terms of post-fertilization time period; structures, and functions observed at the end of this period, using the chick embryo as the organism example.

5. Describe the changes during fetal development. Recognize the **chorion, amnion, and yolk sac in the fetal models.** What are differences between fraternal and identical twins?

6. Describe labor and parturition in terms of onset and phases.

# ACTIVITY 1: IDENTIFY THE PRINCIPLE STAGES OF DEVELOPMENT FROM FERTILIZATION TO UTERINE IMPLANTATION

*You will need starfish embryo slides and a compound light microscope.*

Finish labeling the diagram of fertilization below. The arrows show the progress of the sperm as it enters the oocyte.

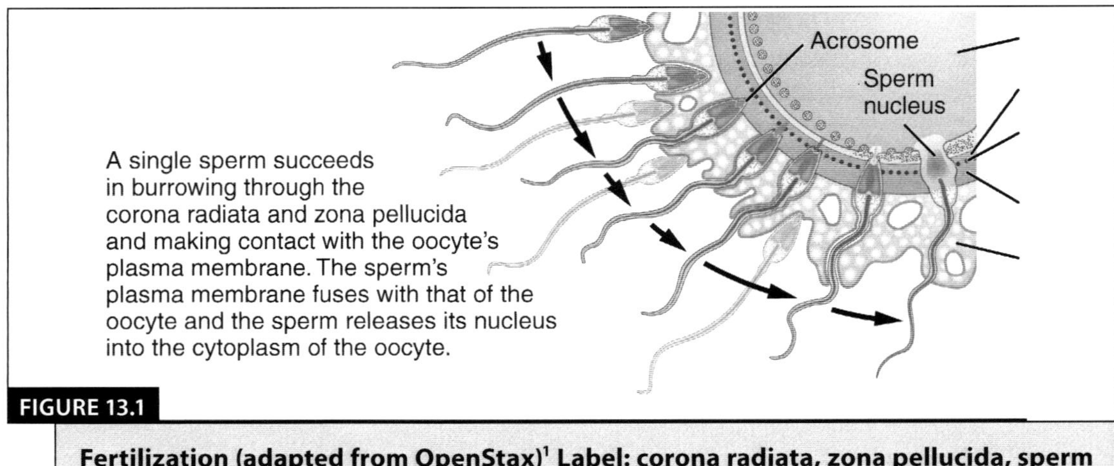


**FIGURE 13.1**

**Fertilization (adapted from OpenStax)[1] Label: corona radiata, zona pellucida, sperm receptors on the plasma membrane, plasma membrane, and cytoplasm**

Fertilization [https://cnx.org/contents/FPtK1zmh@8.108:mg9JPatU@3/Fertilization] by OpenStax, (CC BY 4.0)

Answer these critical thinking questions after reading the online text content and the Pre-Lab information:

1. What is the function of the acrosome?

2. What are the relative contributions of sperm and ovum cytoplasm to the zygote? What is a reason that both males and females inherit their mother's but not their father's mitochondrial DNA?

**3.** What are some reasons that fertilization normally occurs only if intercourse occurs between 3–5 days prior to ovulation and 1 day after ovulation?

**4.** What are some reasons that fertilization normally takes place in the first third of the oviduct—the third nearest the ovary?

5. Using a microscope and slides of starfish embryos, complete the table below, drawing and labeling the indicated developmental stages.

| STAGE | DRAWING | RELATIVE SIZE OF CELLS | NUCLEUS VISIBLE | FERTILIZATION MEMBRANE VISIBLE |
|---|---|---|---|---|
| Ovum | | Large (100 um) | Yes | |
| Zygote | | | No | |
| 4 cell | | | | |
| Morula | | | | |
| Blastula | | | | |

6. Based on your observations, why do you think that cell divisions occurring between fertilization and blastocyst stages are called reduction divisions?

7. What mammalian structure is analogous to the starfish fertilization membrane?

8. What is the function of the fertilization membrane?

## ACTIVITY 2: IDENTIFY MAJOR CHANGES IN THE CHICK EMBRYO OCCURRING DURING THE EMBRYONIC PERIOD ——

*You will need chick embryo slides and a compound light microscope or a dissecting microscope.*

1.  Draw a picture of the earliest stage chick embryo available (27–29 hr) for comparison with the latest stage embryo available (80 hr).

**Early Stage Chick Embryo**                    **Late Stage Chick Embryo**

2.  Complete the table below, indicating which structures have developed, based on your observations of the chick embryos.

| AGE | SOMITES | HEART | HEAD AND NECK STRUCTURES (EYE, GILL SLITS, BRAIN, ETC.) | NEURAL TUBE | LIMB BUDS |
|---|---|---|---|---|---|
| Early | | | | | |
| Late | | | | | |

3.  What is the function of the allantois in the chick embryo? Think of the chorionic villi in the mammal. (The chorion and allantois combine in the chick. The allantois also stores wastes/urine once the kidneys develop.)

## ACTIVITY 3: IDENTIFY FETAL MEMBRANES AND VARIATIONS IN FETAL POSITION WITHIN THE UTERUS IN SINGLE AND MULTIPLE BIRTHS

*You will need models of **human fetal development.***

Answer the following questions based on your knowledge of embryonic development while observing and learning the fetal models in lab. The allantois, chorion, and amnion identify mammals, reptiles, and birds as "amniotes."

1. Where is the chorion in relation to the amnion?

2. From what germ layer does the yolk sac develop? Can you see it on the fetal models?

3. Which membrane or layer(s) contributes to the formation of the umbilical cord? Find the cord on the models.

4. Do fraternal twins share identical genes? Why or why not?

5. At what embryological stage do identical twins first appear?

6. Why are identical twins called "monozygotic" and fraternal twins called "dizygotic?"

7. Draw the following developmental stages on this page below:
   - Fetus in the uterus in vertex position
   - Fetus in the uterus in breech position
   - Fraternal twins each with their own placenta
   - Identical twins sharing the same amnion and placenta

Describe and be able to identify the following cells and embryonic or fetal structures on **models** in the lab. "Fxn" indicates to include function with the description.

1. Secondary oocyte
2. Ovum
3. Sperm (fxn)
4. Corona radiata
5. Zona pellucida
6. Zygote
7. Morula
8. Blastocyst (fxn)
9. Inner cell mass
10. Trophoblast (fxn)
11. Amnion (fxn)
12. Chorion (fxn)
13. Yolk sac
14. Placenta (fxn)
15. Umbilical cord, including arteries and veins (fxn)
16. Chorionic villi (fxn)

Authored by James Hammarback

Reference: OpenStax College. (2013). *Anatomy & Physiology.* Houston, TX: OpenStax CNX.
Retrieved from https://openstax.org/details/books/anatomy-and-physiology

# 14

# INHERITANCE
## PRE-LAB

Name: ________________________  Section: __________  Date: ________

## LEARNING OBJECTIVES

1. Define the basic principles of inheritance.

2. Using a Punnett square, successfully complete and analyze monohybrid crosses with the following inheritance patterns: autosomal dominant-recessive, incomplete dominance, co-dominance, and sex-linked traits.

*Checklist to complete* **before entering** *the science skills lab (SSL):*

☐ Actively read this packet of information.

☐ Complete the charts, tables or labeling and answer questions using your own words.

☐ Complete the electronic digital pre-lab quiz on Bb by Sunday.

☐ Review the attached anatomy list and take it to lab with you, Jot down key descriptive identifying words that aid in your lab test preparations.

## ACTIVITY 1: DEFINE THE BASIC PRINCIPLES OF INHERITANCE AND IDENTIFY DISORDERS ASSOCIATED WITH INHERITANCE

*Read the content in your text and the introductory paragraphs below. Then, complete Activity 1 in lab.*

Each human body cell has a full complement of DNA stored in 23 pairs of chromosomes. Among these is one pair of chromosomes, called the sex chromosomes, that determines the sex of the individual (XX in females, XY in males). The remaining 22 chromosome pairs are called autosomal chromosomes. Each of these chromosomes carries hundreds or even thousands of genes, each of which codes for the assembly of a particular

protein—that is, genes are "expressed" as proteins. An individual's complete genetic makeup is referred to as his or her **genotype.** The characteristics that the genes express, whether they are physical, behavioral, or biochemical, are a person's **phenotype.**

You inherit one chromosome in each pair—a full complement of 23—from each parent. Homologous chromosomes—those that make up a complementary pair—have genes for the same characteristics in the same location on the chromosome. Because one copy of a gene, an **allele,** is inherited from each parent, the alleles in these complementary pairs may vary. Take for example an allele that encodes for dimples. A child may inherit the allele encoding for dimples on the chromosome from the father and the allele that encodes for smooth skin (no dimples) on the chromosome from the mother.

Although a person can have two identical alleles for a single gene (a **homozygous** state), it is also possible for a person to have two different alleles (a **heterozygous** state). The two alleles can interact in several different ways. The expression of an allele can be **dominant,** for which the activity of this gene will mask the expression of a nondominant, or **recessive,** allele. Sometimes dominance is complete; at other times, it is incomplete.

## MENDEL'S THEORY OF INHERITANCE

Our contemporary understanding of genetics rests on the work of a nineteenth-century monk. Working in the mid-1800s, long before anyone knew about genes or chromosomes, Gregor Mendel discovered that garden peas transmit their physical characteristics to subsequent generations in a discrete and predictable fashion. When he mated, or crossed, two pure-breeding pea plants that differed by a certain characteristic, the first-generation offspring all looked like one of the parents. For instance, when he crossed tall and dwarf pure-breeding pea plants, all of the offspring were tall. Mendel called tallness dominant because it was expressed in offspring when it was present in a purebred parent. He called dwarfism recessive because it was masked in the offspring if one of the purebred parents possessed the dominant characteristic. Note that tallness and dwarfism are variations on the characteristic of height. Mendel called such a variation a **trait.** We now know that these traits are the expression of different alleles of the gene encoding height.

It is common practice in genetics to use capital and lowercase letters to represent dominant and recessive alleles. Using Mendel's pea plants as an example, if a tall pea plant is homozygous, it will possess two tall alleles ($TT$). A dwarf pea plant must be homozygous because its dwarfism can only be expressed when two recessive alleles are present (**tt**). A heterozygous pea plant (**Tt**) would be tall and phenotypically indistinguishable from a tall homozygous pea plant because of the dominant tall allele. Because of the random segregation of gametes, the laws of chance and probability come into play when predicting the likelihood of a given phenotype. These probabilities can be easily determined with the use of a **Punnett square.**

The steps to implementing a Punnett square and solving monohybrid crosses are:

1.  Draw a Punnett square.

2.  Separate the alleles from each parent and place them on the boundaries of the Punnett square (see the black arrows).

3.  Complete the Punnett square by bringing the allele down and over (see the dashed gray arrows).

4.  Answer the questions accompanying the cross problem by determining probabilities.

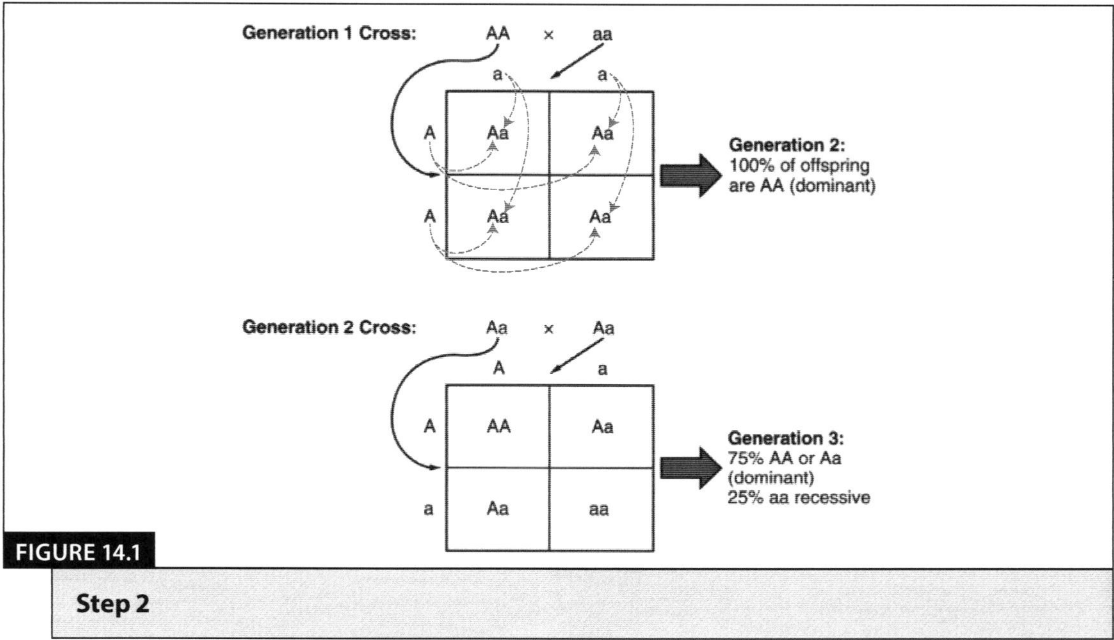


**FIGURE 14.1**

**Step 2**

Complete the following chart.

| TERM | DEFINITION | EXAMPLE |
| --- | --- | --- |
| Genotype | | |
| Phenotype | | |
| Allele | | |
| Homozygous | | |
| Heterozygous Dominant | | |
| Homozygous Recessive | | |

## ACTIVITY 2: USING A PUNNETT SQUARE, SUCCESSFULLY COMPLETE AND ANALYZE MONOHYBRID CROSSES WITH THE FOLLOWING INHERITANCE PATTERNS: COMPLETE DOMINANCE, INCOMPLETE DOMINANCE, CO-DOMINANCE, AND SEX-LINKED TRAITS

*Read the content in your text and the introductory paragraphs below. Then, complete Activity 2 in lab.*

### AUTOSOMAL DOMINANT-RECESSIVE INHERITANCE

When a dominant allele is located on one of the 22 pairs of autosomes (non-sex chromosomes), we refer to its inheritance pattern as autosomal dominant. When a genetic disorder is inherited in an autosomal recessive pattern, the disorder corresponds to the homozygous recessive phenotype. With this pattern of inheritance, the dominant allele will completely mask the expression of the recessive counterpart. This means that heterozygous individuals will not display symptoms of this disorder because their unaffected gene will compensate. Such an individual is called a carrier. Carriers for an autosomal recessive disorder may never know their genotype unless they have a child with the disorder.

Let's use cystic fibrosis (CF) as our example. CF is characterized by the chronic accumulation of a thick, tenacious mucus in the lungs and digestive tract. CF is a relatively common disorder that occurs in approximately 1 in 2000 Caucasians. A child born to two CF carriers would have a 25 percent chance of inheriting the disease. The pattern and corresponding Punnett square is shown below.

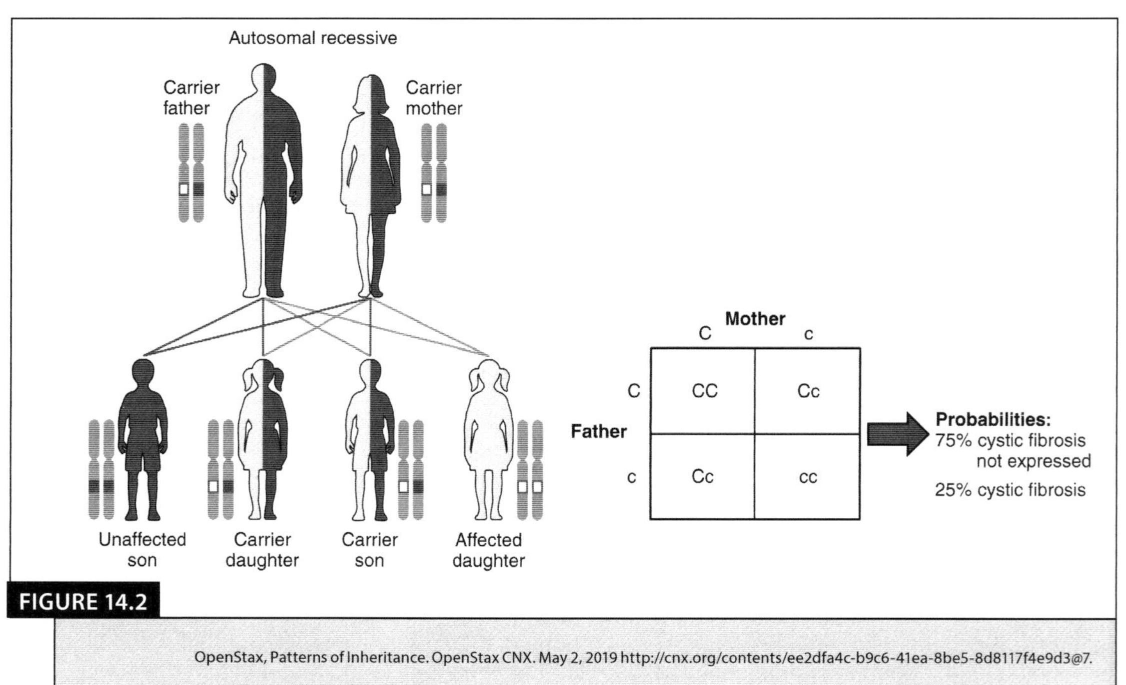


**FIGURE 14.2**

OpenStax, Patterns of Inheritance. OpenStax CNX. May 2, 2019 http://cnx.org/contents/ee2dfa4c-b9c6-41ea-8be5-8d8117f4e9d3@7.

**1.** A child born to a CF carrier and someone with two unaffected alleles would have a _______ percent probability of inheriting CF; this same child would have a _______ percent chance of being a carrier.

## INCOMPLETE DOMINANCE

In incomplete dominance, the offspring express a heterozygous phenotype that is intermediate between one parent's homozygous dominant trait and the other parent's homozygous recessive trait. In this case, three alleles exist for a trait instead of only two. All three alleles are written with lowercase letters and create a blended intermediate phenotype when mixed. Let's look at one of the genes responsible for hair texture. When one parent (cc) passes a curly hair allele (c) and the other parent (ss) passes a straight-hair allele (s), the effect on the offspring will be intermediate, resulting in hair that is wavy (cs).

|   | c | c |
|---|---|---|
| s | cs | cs |
| s | cs | cs |

**1.** What is the phenotype of the offspring produced from crossing red flowers with white flowers, assuming the same form of inheritance is followed?

## CO-DOMINANCE

Co-dominance is characterized by the equal, distinct, and simultaneous expression of both parents' different alleles. This pattern differs from the intermediate, blended features seen in incomplete dominance. A classic example of codominance in humans is ABO blood type. People are blood type A if they have an allele for an enzyme that facilitates the production of surface antigen A on their erythrocytes. This allele is designated $I^A$. In the same manner, people are blood type B if they express an enzyme for the production of surface antigen B. People who have alleles for both enzymes produce both surface antigens A and B. As a result, they are blood type AB. Because the effect of both alleles (and enzymes) is observed, we say that the $I^A$ and $I^B$ alleles are co-dominant.

There is also a third allele that determines blood type. This allele (*i*) produces a nonfunctional enzyme. People who have two *i* alleles do not produce either A or B surface antigens; they have type O blood. If a person has *A* and *i* alleles, the person will have blood type A. Notice that it does not make any difference whether a person has two $I^A$ alleles or one $I^A$ and one *i* allele. In both cases, the person is blood type A. Because $I^A$ masks *i*, we say that $I^A$ is dominant to *i*.

Complete the table below.

| EXPRESSION OF BLOOD TYPES | | |
|---|---|---|
| **BLOOD TYPE** | **GENOTYPE** | **PHENOTYPE** |
| A | _____ or _____ | |
| B | _____ or _____ | |
| AB | | |
| O | | |

Let's use the example that a woman with type AB blood mates with a man that has type O blood. We know that the alleles for the woman are $I^A$ and $I^B$ while the alleles for the male are *i and i*. This means …

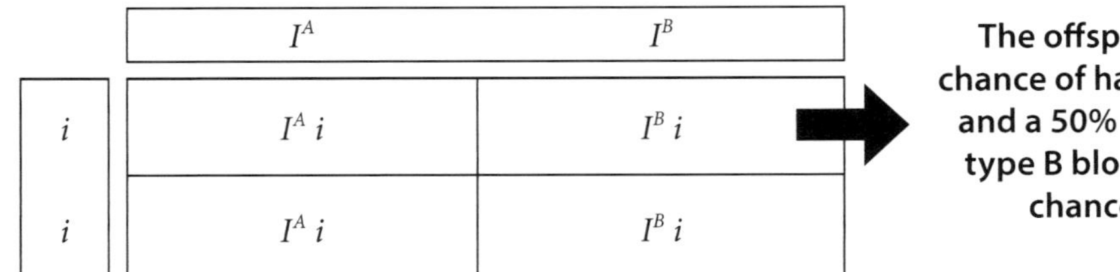


## SEX-LINKED TRAITS

An X-linked transmission pattern involves genes located on the X chromosome of the 23rd pair. Recall that a male has one X and one Y chromosome. When a father transmits a Y chromosome, the child is male, and when he transmits an X chromosome, the child is female. A mother can transmit only an X chromosome, as both her sex chromosomes are X chromosomes.

When an abnormal allele for a gene that occurs on the X chromosome is dominant over the normal allele, the pattern is described as X-linked dominant. This is the case with vitamin D—resistant Rickets; an affected father would pass the disease gene to all of his daughters, but none of his sons because he donates only the Y chromosome to his sons. For an affected female, the inheritance pattern would be identical to that of an autosomal dominant inheritance pattern in which one parent is heterozygous, and the other is homozygous for the normal gene.

X-linked recessive inheritance is much more common because females can be carriers of the disease yet still have a normal phenotype. With X-linked recessive diseases, males either have the disease or are genotypically normal—they cannot be carriers. Females, however, can be genotypically normal, a carrier who is phenotypically normal, or affected with the disease. For an example of X-linked recessive inheritance, consider parents in which the mother is an unaffected carrier ($X^N X^n$) and the father is normal ($X^N Y$). We are using N = normal, unaffected and n = disease, affected. None of the daughters would have the disease because they receive a normal gene from their father. However, they have a 50 percent chance of receiving the disease gene from their mother and becoming a carrier. In contrast, 50 percent of the sons would be affected.

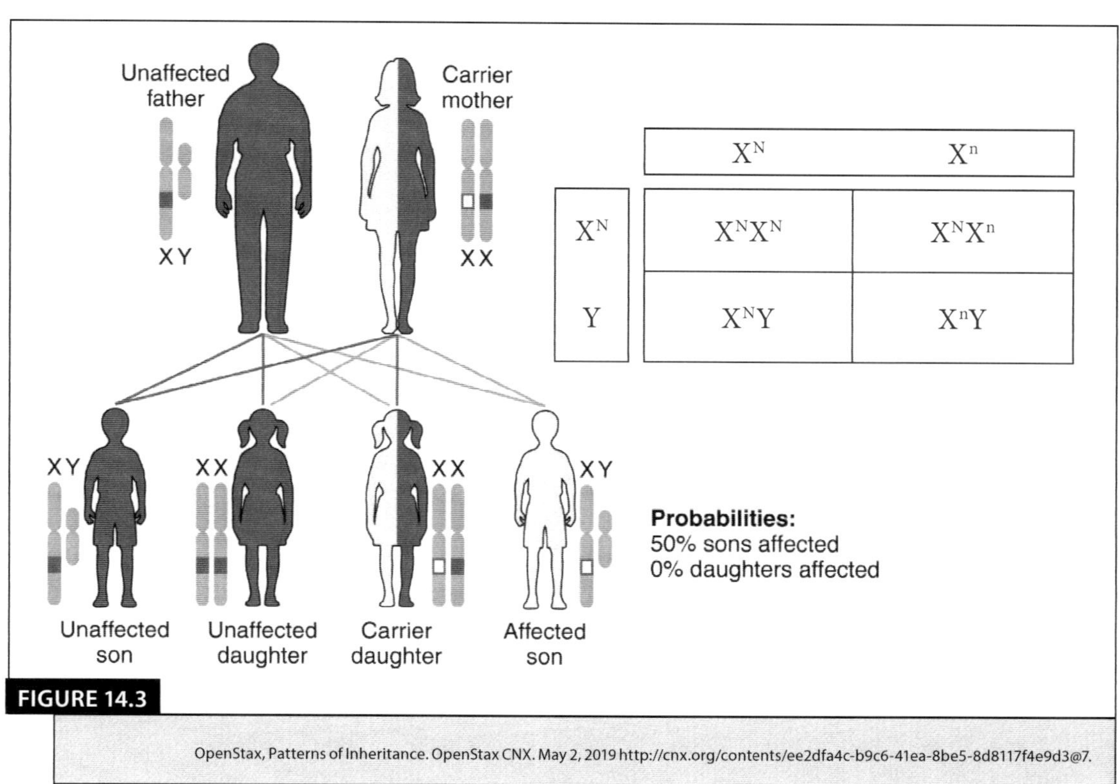


|  | $X^N$ | $X^n$ |
|---|---|---|
| $X^N$ | $X^N X^N$ | $X^N X^n$ |
| Y | $X^N Y$ | $X^n Y$ |

**FIGURE 14.3**

OpenStax, Patterns of Inheritance. OpenStax CNX. May 2, 2019 http://cnx.org/contents/ee2dfa4c-b9c6-41ea-8be5-8d8117f4e9d3@7.

1. Do X-linked recessive disorders affect many more males than females? Why?

2. How can a female carrier of an X-linked recessive disorder have a daughter who is affected?

*Checklist to complete* **before entering** *the science skills lab (SSL):*

☐ Actively read this packet of information.

☐ Complete the charts, tables or labeling and answer questions using your own words.

☐ Complete the electronic digital pre-lab quiz on Bb by Sunday.

☐ Review the attached anatomy list and take it to lab with you. Jot down key descriptive identifying words that aid in your lab test preparations.

## ANATOMY LIST

1. Definitions:
   a. Genotype
   b. Phenotype
   c. Allele
   d. Homozygous
   e. Heterozygous
   f. Dominant
   g. Recessive
   h. Trait
   i. Punnett square

2. Solve monohybrid cross problems for the following types of inheritance:
   a. Autosomal dominant-recessive
   b. Incomplete dominance
   c. Co-dominance
   d. Sex-linked

# 14

# INHERITANCE
## IN-LAB ACTIVITIES

Name: ___________________________     Section: ___________     Date: _________

## LEARNING OBJECTIVES

1. Define the basic principles of inheritance.

2. Using a Punnett square, successfully complete and analyze monohybrid crosses with the following inheritance patterns: autosomal dominant-recessive, incomplete dominance, co-dominance, and sex-linked traits.

## ACTIVITY 1: DEFINE THE BASIC PRINCIPLES OF INHERITANCE

1. Name the genotype next to each allelic pairing below.

| | |
|---|---|
| BB | |
| Hh | |
| ff | |
| Dd | |
| LL | |
| aa | |

2. Name the phenotype next to each allelic pairing below. Let B = brown eyes and b = blue eyes.

| | |
|---|---|
| BB | |
| Bb | |
| bb | |

3. Review the steps to implementing a Punnett square and solving the monohybrid cross shown. A man who is heterozygous for eye color (Bb) mates with a woman with blue eyes (bb).

4. ▪ Draw a Punnett square.

▪ Separate the alleles from each parent and place them on the boundaries of the Punnett square. (see below)

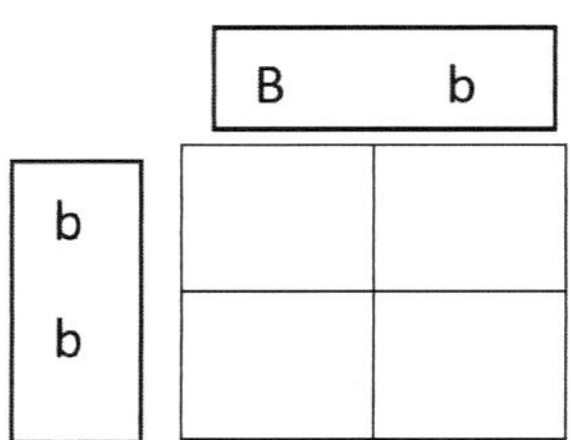


▪ Complete the Punnett square by bringing the alleles down and over.

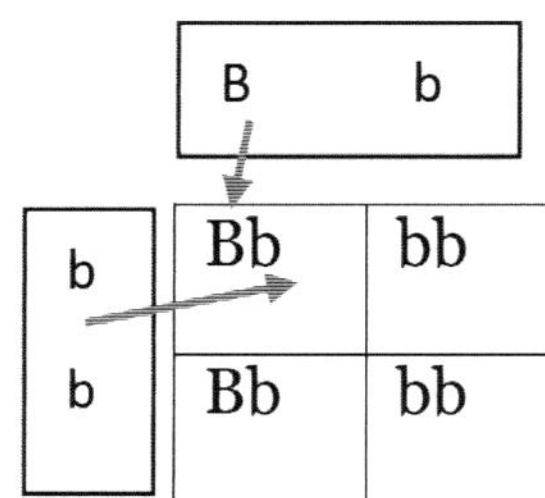


▪ Answer the questions accompanying the cross problem by determining probabilities.

## ACTIVITY 2: USING A PUNNETT SQUARE, SUCCESSFULLY COMPLETE AND ANALYZE MONOHYBRID CROSSES WITH THE FOLLOWING INHERITANCE PATTERNS: AUTOSOMAL DOMINANT- RECESSIVE, INCOMPLETE DOMINANCE, CO-DOMINANCE, AND SEX-LINKED TRAITS.

### AUTOSOMAL DOMINANT–RECESSIVE INHERITANCE

Solve the following problems:

1. The presence of dimples is considered a dominant trait so let D represent the presence of dimples and d represent the absence of dimples. The male and female are both heterozygous for dimples.

   **a.** Draw and solve a Punnett square for this problem.

   **b.** What % of the offspring have dimples?

   _______________________________________________

   **c.** What % of the offspring have the genotype Dd?

   _______________________________________________

2. Let R = the ability to roll ones tongue into a "U" shape and r = the inability to roll the tongue. The male is homozygous dominant while the female is homozygous recessive.

   **a.** Draw and solve a Punnett square for this problem.

    **b.** What % of the offspring lack the ability to roll their tongue?

_______________________________________________________________

    **c.** What % of the offspring have the genotype RR?

_______________________________________________________________

**3.** W= widow's peak hairline and w=straight hairline. The male has a widow's peak while the female does not.

    **a.** Draw and solve a Punnett square for this problem. ***Hint: you need to draw 2.***

    **b.** Can any of the combinations produce offspring with a straight hairline?

_______________________________________________________________

**4.** F = free ear lobes and f = attached ear lobes. The male has attached ear lobes while the female is heterozygous.

    **a.** Draw and solve a Punnett square for this problem.

    **b.** What % of the offspring have attached earlobes?

_______________________________________________________________

    **c.** What % of the offspring have the genotype Ff?

_______________________________________________________________

## INCOMPLETE DOMINANCE

Solve the following problems:

1. Let cc = curly hair, sc = wavy hair, and ss= straight hair. The male has curly hair and the female has wavy hair.

    **a.** Draw and solve a Punnett square for this problem.

    **b.** What % of the offspring have curly hair?

    _______________________________________________________________

    **c.** What % of the offspring have the genotype ss?

    _______________________________________________________________

    **d.** What % of the offspring have wavy hair?

    _______________________________________________________________

2. Let ww = white flower color, wr = pink flower color, and rr = red flower color. The male has white flowers and the female has red flowers.

    **a.** Draw and solve a Punnett square for this problem.

    **b.** What is the genotype of all of the offspring?

    _______________________________________________________________

    **c.** What is the phenotype of all of the offspring?

    _______________________________________________________________

3. Use the same flower color alleles from problem 2. The male and female both have pink flowers.

   a. Draw and solve a Punnett square for this problem.

   b. What is the genotypic ratio of the offspring?

   c. What % of the offspring have white flowers?

## CO-DOMINANCE

Solve the following problems:

1. Let AA and AO = type A blood, BB and BO = type B blood, AB = type AB blood, and OO = type O blood. The male has type O blood and the female has type AB blood.

   a. Draw and solve a Punnett square for this problem.

   b. What % of the offspring have type O blood?

   c. What % of the offspring have the genotype AO?

   d. What % of the offspring have the genotype AB?

2. Use the same blood type alleles from problem 1. The male has type A blood and the female has type B blood.

   a. Draw and solve a Punnett square for this problem. *Hint: you need to draw 4.*

   b. Can any of the combinations produce offspring with type O blood?

   c. Can any of the combinations produce offspring with type AB blood?

3. Use the same blood type alleles from problem 1. The male and female both have type AB blood.

   a. Draw and solve a Punnett square for this problem.

   b. What % of the offspring have a type B blood?

   c. What % of the offspring have the genotype AB?

## SEX—LINKED RECESSIVE INHERITANCE

Solve the following problems:

1.  Sickle cell is a sex-linked recessive disorder carried on the X chromosome. The male is unaffected ($X^N Y$) and the female is a heterozygous carrier ($X^N X^n$).

    **a.** Draw and solve a Punnett square for this problem.

    **b.** What % of the offspring will have sickle cell/affected?

    _______________________________________________________________________

    **c.** What % of the MALE offspring will have sickle cell/affected?

    _______________________________________________________________________

2.  Color blindness is a sex-linked recessive disorder carried on the X chromosome. The male is color blind/affected ($X^n Y$) and the female is heterozygous/unaffected ($X^N X^n$).

    **a.** Draw and solve a Punnett square for this problem.

    **b.** What % of the offspring will be color blind/affected?

    _______________________________________________________________________

    **c.** What % of the FEMALE offspring will be a carrier for this allele?

    _______________________________________________________________________

3. Baldness is a sex-linked recessive disorder carried on the X chromosome. What parental genotypes will produce bald FEMALE offspring?

   a. Draw and solve a Punnett square for this problem. ***Hint: you need to draw 6.*** Circle the ones that produce bald FEMALE offspring.

4. Hemophilia is a sex-linked recessive disorder carried on the X chromosome. Can a female with hemophilia have a son that does NOT have hemophilia? Why or why not?

_______________________________________________________________________

_______________________________________________________________________

_______________________________________________________________________

_______________________________________________________________________

_______________________________________________________________________

# 15

# CUMULATIVE REVIEW
## IN-LAB ACTIVITIES

Name: _________________________  Section: __________  Date: _________

## ACTIVITY 1: COMPLETE THE CUMULATIVE REVIEW SHEET IN PREPARATION FOR THE UPCOMING CUMULATIVE FINAL EXAM

*You will need all of the course materials you collected and brought with you to lab today.*

In small groups (recommended), attempt to answer as many questions as you can. You may **only** use the materials that either you or your lab partners brought with you to lab today—excluding the internet (use each other's memory and **not** the book or Internet sources, so that you can discern what you have retained and not retained from A&P **before** the test). This review assignment will help to identify material that you must review before the upcoming final.

1. Contrast hydrophilic and hydrophobic hormones. Include in your answer: how they are transported in the blood, where the receptor on the target cell is located, and examples of each type of hormone.

______________________________________________________

______________________________________________________

2. A patient has elevated GnRH levels, low FSH and LH levels, and low testosterone levels. In which tissue is the pathology located? Is this a primary or secondary pathology?

______________________________________________________

______________________________________________________

______________________________________________________

3. Name, in order from superficial to deep, all of the layers of the heart wall–make sure to include alternate names for any of the layers.

4. What dictates blood flow?

5. Why is the SA node considered to be the pacemaker of the heart?

6. What events does each part of an EKG represent?

7. List the four cardinal signs of inflammation.

8. Distinguish between natural passive, natural active, artificial passive, and artificial active immunity. Give examples of each.

9. What is the significance of the glottis?

10. Draw a hemoglobin saturation curve. Explain the 5 saturation at $pO_2$ of 100 and $pO_2$ of 40. Then, list and explain all of the factors that shift the curve left and right (6 factors for each) and what it means when the curve shifts.

11. Draw how $O_2$ enters and leaves the blood. Include in your answer how $O_2$ is transported in the blood. (Include numerical values.)

12. Do the same for $CO_2$.

**13.** List three modifications of the small intestine that increase surface area for absorption.

**14.** List three filtration barriers in the renal corpuscle.

**15.** Reconstruct the urinary flow chart including blood flow, filtrate flow, and urine flow.

**16.** Define the following:

   **a.** Filtration

   **b.** Reabsorption

   **c.** Secretion

   **d.** Excretion

**17.** Explain the RAAS pathway.

**18.** Calculate TBW, ECF volume, and ICF volume from a 70 kg man.

**19.** What are the structures that regulate testicular temperature, and how do they do it?

**20.** Give the five hormones of the male reproductive system and their function.

**21.** Give the eight hormones of the female reproductive system and their function (think flow chart).

**22.** Where does fertilization usually happen?

__________________________________________________

__________________________________________________

__________________________________________________

**23.** For the following give their jobs:

    **a.** IgA

__________________________________________________

    **b.** IgG

__________________________________________________

    **c.** IgM

__________________________________________________

    **d.** IgE

__________________________________________________

    **e.** IgD

__________________________________________________

**24.** Define the following respiratory volumes:

    **a.** TV

__________________________________________________

    **b.** IRV

__________________________________________________

    **c.** ERV

__________________________________________________

    **d.** VC

__________________________________________________

    **e.** TLC

__________________________________________________

**25.** Detail the four step process of cellular respiration.

**26.** How during quiet inhalation … _________________ muscle contracts, thoracic volume _________________ , lung pressure _________________ , air flows _________________ , and the lungs _________________ .

**27.** What muscle contracts to cause quiet exhalation?

**28.** Give at least five homologous structures that exist between the male and female reproductive system.

**29.** What is the placenta, and what is its job?

**30.** List all pituitary hormones (all 9) and what their functions are.

**31.** Diagram blood flow through the heart and body (in a big familiar flow chart).

**32.** Review ECG strips and be able to calculate a patient's heart rate. Discuss if it is in the normal range, label all waveforms, and identify abnormalities (elevated waveforms, AV blocks, tachycardia, bradycardia, atrial fibrillation, and ventricular fibrillation)

**33.** List three things that cause acidosis.

**34.** List each part of the nephron and what activities occur in that section.

**35.** In fruit flies, eye color is a sex-linked trait carried on the X chromosome. Let R = red eyes and r = white eyes. The male is heterozygous and the female has white eyes. Predict the resulting offspring.

**36.** Jan visited the doctor after she lost weight, developed a tremor, and developed tachycardia. She also reports that she feels very hot and sweaty almost all the time, even in a cooler room. The doctors discover that Jan has a thyroxine secreting tumor on her thyroid.

   **a.** Would her TRH be elevated or low? Why?

   **b.** Would her TSH be elevated or low? Why?

   **c.** Would her T3/T4 be elevated or low? Why?

**37.** List below each type of formed element in the blood and its function (7).

**38.** A patient comes in with a blood type of A– and needs a blood transfusion. List all compatible blood types (include any + or –).

**39.** Compare and contrast innate immunity versus adaptive immunity. Be detailed.

**40.** Diagram spermatogenesis and oogenesis below.

**41.** At a somatic capillary, describe all Starling forces at work. It may be helpful to diagram this out.

**42.** Describe the boss—factory relationship between the adenohypophysis and the hypothalamus. Be specific. It may help to diagram this out.

**43.** Explain the difference between ventilation and respiration. Include the difference between all three types of respiration.

**44.** Review all hormones of the endocrine system.